Techniques in
Agricultural Microbiology

Techniques in Agricultural Microbiology

Dr. Awani Kr. Singh

Editor

KOROS PRESS LIMITED

London, UK

Techniques in Agricultural Microbiology

© 2012

Printed in 2017 for Sale in the Indian Subcontinent

Published by
Koros Press Limited
3 The Pines, Rubery B45 9FF, Rednal,
Birmingham, United Kingdom

Tel.: +44-7826-930152
Email: info@korospress.com
www.korospress.com

ISBN: 978-1-78163-003-7
Editor: Dr. Awani Kr. Singh

10 9 8 7 6 5 4 3 2 1

British Library Cataloguing in Publication Data
A CIP record for this book is available from the British Library

Exclusively distributed by CBS Publishers & Distributors Pvt. Ltd.
Sales & Distribution Rights only for India, Pakistan, Bangladesh, Sri Lanka, Nepal and Bhutan.This book is not to be sold outside these territories.

Contents

Preface   *(vii)*

1. **Agricultural Microbes**   1
   Virus • Microbiology • Viruses and Human Disease • Bacteria
   • History of Bacteriology • Cellular Structure • Metabolism
   • Interactions with other Organisms • Different Types of
   Bacteria • Fungus • Unique Features • Morphology • Other
   Sexual Processes • Fungus-like Organisms • Edible and
   Poisonous Species • Fungal Endophytes: Common Host Plant
   Symbionts but Uncommon Mutualists • Methods • Results
   • Discussion • Conclusions

2. **Microbiology in Biotechnology**   75
   Microbiology in Biotechnology • Microbes and Agriculture
   • Microbes and Public Health • Microbes in Recovery of Metals
   • Biotechnology in Agriculture : The New Green Revolution
   • Protein Synthesis • Southern and Northern Blot Techniques
   • Recombinant DNA Technology and Gene Cloning • Gene
   Exchange in Bacteria • The Agrobacterium Mediated Transfer
   of Genes • Procedures for Agrobacterium Mediated
   Transformations • Direct DNA-Transfer Technologies
   • Antisense RNA Strategy • Frost Control Biotechnology • Virus
   Resistance in Transgenic Plants • Transformation of Sequences
   Related to Viral Capsid Protein • Transformation of Sequences
   Related to Viral Movement Protein • Resistance Conferred by
   Antisense RNA • Antibody-Mediated Resistance

3. **Role of Bacteria and Actinomycetes in
   Biological Control of Pests and Diseases**   100
   The Location, Living Conditions, and Functions of
   Microorganisms • Soil Bacteria • Mode of Entry of Microbial
   Pathogens on Insect Pests and their Mechanism of Control
   • Identification of Active Principles • Entomopathogens for the
   Control of Storage Pests • Insect Resistance : Its Impact on
   Microbial Control of Insect Pests • Insects vs. Vertebrate

Immunity · Non-specific Defence Mechanisms · Specific Defence Mechanisms/Immunity

4. **DNA Vaccination**   125
Current Use · Plasmid Vectors for Use in Vaccination · Delivery Methods · Immune Response Raised by DNA Vaccines · Mechanistic Basis for DNA Raised Immune Responses · Modulation of the Immune Response · Additional Methods of Enhancing DNA-Raised Immune Responses · Vector DNA · HIV Vaccine

5. **Biological Warfare : A Serious Threat in the 21st Century**   150
Characteristics of Biological Weapons · Biodefence of Troops in the Field · The Biological Weapons Convention · List of BW Institutions and Programs by Country · Alleged Uses · Experimentation and Testing · Experiments on Non-consenting Individuals · Biological Weapons · Notable Outbreaks and Accidents · Sverdlovsk Anthrax Leak · Poison Laboratory of the Soviet Secret Services · Vozrozhdeniya · The Programme · AIDS Origins Conspiracy Theories · Antibiotic Resistance · Resistant Pathogens · Biological Hazard · Levels of Biohazard · Biosecurity · Animal Biosecurity · The Role of Education in Biosecurity · Agents of Concern · Entomological Warfare · World War II · Ethnic Bioweapon · Deinococcus Radiodurans: A Marvelous Berry that Withstands Radiations · Name and Classification · Radiotrophic Fungus

6. **Relation of Bacteria to Agriculture**   230
The Silo · The Fertility of The Soil · Sustainable Agriculture Focus on Issues Facing Farmers and Producers · Soil Health · Decline of Kansas Soils · Nitrogen and Organic Matter · Soil Bacteria · Analysis of the Soil · Western Kansas Cropping Systems · Organic Matter Content of Western Kansas Soils · Looking to the Future of Kansas Soils · Role of Bacteria in Coffee Plantation Ecology · Distribution and Functions of Bacteria · Environmental Factors · Nutritional Requirements of Bacteria: Macronutrients · Specialised Bacterial Cells · Most Commonly Encountered Soil Bacteria · Conclusion

Bibliography   266

Index   270

Preface

Viruses display a wide diversity of shapes and sizes, called *morphologies*. Viruses are about 1/100th the size of bacteria. Most viruses that have been studied have a diameter between 10 and 300 nanometres. Some filoviruses have a total length of up to 1400 nm, however their diameters are only about 80 nm. Most viruses are unable to be seen with a light microscope so scanning and transmission electron microscopes are used to visualise virus particles. To increase the contrast between viruses and the background, electron-dense "stains" are used. These are solutions of salts of heavy metals such as tungsten, that scatter the electrons from regions covered with the stain. When virus particles are coated with stain, fine detail is obscured. Negative staining overcomes this problem by staining the background only.

A complete virus particle, known as a virion, consists of nucleic acid surrounded by a protective coat of protein called a capsid. These are formed from identical protein subunits called capsomers. Viruses can have a lipid "envelope" derived from the host cell membrane. The capsid is made from proteins encoded by the viral genome and its shape serves as the basis for morphological distinction. Virally coded protein subunits will self-assemble to form a capsid, generally requiring the presence of the virus genome. However, complex viruses code for proteins that assist in the construction of their capsid. Proteins associated with nucleic acid are known as nucleoproteins, and the association of viral capsid proteins with viral nucleic acid is called a nucleocapsid. The capsid and entire virus structure can be mechanically (physically) probed through atomic force microscopy. In general, there are four main morphological virus types:

These viruses are composed of a single type of capsomer stacked around a central axis to form a helical structure, which may have a central cavity, or hollow tube. This arrangement results in rod-shaped or filamentous virions: these can be short and highly rigid, or long and very flexible. The genetic material, generally single-stranded

RNA, but ssDNA in some cases, is bound into the protein helix, by interactions between the negatively charged nucleic acid and positive charges on the protein. Overall, the length of a helical capsid is related to the length of the nucleic acid contained within it and the diameter is dependent on the size and arrangement of capsomers. The well-studied Tobacco mosaic virus is an example of a helical virus.

It is our hope that this book will supply to the needs and future prospects of all agricultural biotechnologies and microbiologies.

—Editor

Agricultural Microbes

Virus

A virus (from the Latin *virus* meaning *toxin* or *poison*) is a microscopic infectious agent that can reproduce only inside a host cell. Viruses infect all types of organisms: from animals and plants, to bacteria and archaea. Since the initial discovery of tobacco mosaic virus by Martinus Beijerinck in 1898, more than 5,000 types of virus have been described in detail, although most types of virus remain undiscovered.

Viruses are ubiquitous, as they are found in almost every ecosystem on Earth, and are the most abundant type of biological entity on the planet. The study of viruses is known as virology, and is a branch of microbiology. Viruses consist of two or three parts: all viruses have genes made from either DNA or RNA, long molecules that carry genetic information; all have a protein coat that protects these genes; and some have an envelope of fat that surrounds them when they are outside a cell.

Viruses vary in shape from simple helical and icosahedral shapes, to more complex structures. They are about 1/100th the size of bacteria. The origins of viruses in the evolutionary history of life are unclear: some may have evolved from plasmids—pieces of DNA that can move between cells—while others may have evolved from bacteria. In evolution, viruses are an important means of horizontal gene transfer, which increases genetic diversity.

Viruses spread in many ways; plant viruses are often transmitted from plant to plant by insects that feed on sap, such as aphids, while animal viruses can be carried by blood-sucking insects. These disease-bearing organisms are known as *vectors*. Influenza viruses are spread by coughing and sneezing, and others such as norovirus, are transmitted by the faecal-oral route, when they contaminate hands, food, or water. Rotaviruses are often spread by direct contact with infected children. HIV is one of several viruses that are transmitted through sexual contact.

Not all viruses cause disease, as many viruses reproduce without causing any obvious harm to the infected organism. Viruses such as hepatitis B can cause life-long or chronic infections, and the viruses continue to replicate in the body despite the hosts' defence mechanisms. In some cases, these chronic infections might be beneficial as they might increase the immune system's response against infection by other pathogens. However, in most cases viral infections in animals cause an immune response that eliminates the infecting virus. These immune responses can also be produced by vaccines that give immunity to a viral infection. Microorganisms such as bacteria also have defences against viral infection, such as restriction modification systems. Antibiotics have no effect on viruses, but antiviral drugs have been developed to treat both life-threatening and more minor infections.

Etymology

The word is from the Latin *virus* referring to poison and other noxious substances, first used in English in 1392. *Virulent*, from Latin *virulentus* (poisonous) dates to 1400. A meaning of "agent that causes infectious disease" is first recorded in 1728, before the discovery of viruses by Dmitry Ivanovsky in 1892. The adjective *viral* dates to 1948. The term *virion* is also used to refer to a single infective viral particle. The plural of virus is "viruses".

History

In 1884, the French microbiologist Charles Chamberland invented a filter (known today as the Chamberland filter or Chamberland-Pasteur filter), with pores smaller than bacteria. Thus, he could pass a solution containing bacteria through the filter and completely remove them from the solution. In 1892 the Russian biologist Dimitri Ivanovski used this filter to study what is now known to be tobacco mosaic virus. His experiments showed that the crushed leaf extracts from infected tobacco plants are still infectious after filtration. Ivanovski suggested the infection might be caused by a toxin produced by bacteria, but did not pursue the idea. At the time it was thought that all infectious agents could be retained by filters and grown on a nutrient medium—this was part of the germ theory of disease. In 1898 the Dutch microbiologist Martinus Beijerinck repeated the experiments and became convinced that this was a new form of infectious agent. He went on to observe that the agent multiplied only in dividing cells, but as his experiments did not show that it was made of particles, he called it a *contagium vivum fluidum* (soluble living germ) and re-introduced the word *virus*. Beijerinck maintained that viruses were liquid in nature, a theory later discredited by Wendell Stanley, who proved they were particulate. In the same year, 1899, Friedrich Loeffler and Frosch passed the agent of foot and mouth disease (aphthovirus) through a similar filter and ruled out the possibility of a toxin because of the high dilution; they concluded that the agent could replicate.

In the early 20th century, the English bacteriologist Frederick Twort discovered the viruses that infect bacteria, which are now called bacteriophages, and the French-Canadian microbiologist Felix d'Herelle described viruses that, when added to bacteria on agar, would produce areas of dead bacteria. He accurately diluted a suspension of these viruses and discovered that the highest dilutions, rather than killing all the bacteria, formed discrete areas of dead organisms. Counting these areas and multiplying by the dilution factor allowed him to calculate the number of viruses in the suspension.

By the end of the nineteenth century, viruses were defined in terms of their infectivity, filterability, and their requirement for living hosts. Viruses had been grown only in plants and animals. In 1906, Harrison invented a method for growing tissue in lymph, and, in 1913, E. Steinhardt, C. Israeli, and R. A. Lambert used this method to grow vaccinia virus in fragments of guinea pig corneal tissue. In 1928, H. B. Maitland and M. C. Maitland grew vaccinia virus in suspensions of minced hens' kidneys. Their method was not widely adopted until the 1950s, when poliovirus was grown on a large scale for vaccine production.

Another breakthrough came in 1931, when the American pathologist Ernest William Goodpasture grew influenza and several other viruses in fertilised chickens' eggs. In 1949 John F. Enders, Thomas Weller, and Frederick Robbins grew polio virus in cultured human embryo cells, the first virus to be grown without using solid animal tissue or eggs. This work enabled Jonas Salk to make an effective polio vaccine.

With the invention of electron microscopy in 1931 by the German engineers Ernst Ruska and Max Knoll came the first images of viruses. In 1935 American biochemist and virologist Wendell Stanley examined the Tobacco mosaic virus and found it to be mostly made from protein. A short time later, this virus was separated into protein and RNA parts. Tobacco mosaic virus was the first one to be crystallised and whose structure could therefore be elucidated in detail.

The first X-ray diffraction pictures of the crystallised virus were obtained by Bernal and Fankuchen in 1941. Based on her pictures, Rosalind Franklin discovered the full structure of the virus in 1955. In the same year, Heinz Fraenkel-Conrat and Robley Williams showed that purified Tobacco mosaic virus RNA and its coat protein can assemble by themselves to form functional viruses, suggesting that this simple mechanism was probably how viruses assembled within their host cells.

The second half of the twentieth century was the golden age of virus discovery and most of the 2,000 recognised species of animal, plant, and bacterial viruses were discovered during these years. In 1957, equine arterivirus and the cause of Bovine virus diarrhea (a pestivirus) were discovered.

In 1963, the hepatitis B virus was discovered by Baruch Blumberg, and in 1965, Howard Temin described the first retrovirus. Reverse transcriptase, the key enzyme that retroviruses use to translate their RNA into DNA, was first described in 1970, independently by Howard Temin and David Baltimore. In 1983 Luc Montagnier's team at the Pasteur Institute in France, first isolated the retrovirus now called HIV.

Origins

Viruses are found wherever there is life and have probably existed since living cells first evolved. The origin of viruses is unclear because they do not form fossils, so molecular techniques have been the most useful means of investigating how they arose. These techniques rely on the availability of ancient viral DNA or RNA, but, unfortunately, most of the viruses that have been preserved and stored in laboratories are less than 90 years old. There are three main hypotheses that try to explain the origins of viruses:

Regressive Hypothesis

Viruses may have once been small cells that parasitised larger cells. Over time, genes not required by their parasitism were lost. The bacteria rickettsia and chlamydia are living cells that, like viruses, can reproduce only inside host cells. They lend support to this hypothesis, as their dependence on parasitism is likely to have caused the loss of genes that enabled them to survive outside a cell. This is also called the *degeneracy hypothesis.*

Cellular Origin Hypothesis

Some viruses may have evolved from bits of DNA or RNA that "escaped" from the genes of a larger organism. The escaped DNA could have come from plasmids (pieces of naked DNA that can move *between* cells) or transposons (molecules of DNA that replicate and move around to different positions *within* the genes of the cell). Once called "jumping genes", transposons are examples of mobile genetic elements and could be the origin of some viruses. They were discovered in maize by Barbara McClintock in 1950. This is sometimes called the *vagrancy hypothesis.*

Coevolution Hypothesis

Viruses may have evolved from complex molecules of protein and nucleic acid at the same time as cells first appeared on earth and would have been dependent on cellular life for many millions of years. Viroids are molecules of RNA that are not classified as viruses because they lack a protein coat. However, they have characteristics that are common to several viruses and are often called subviral agents. Viroids are important pathogens of plants. They do not code for proteins but interact with the host cell and use the host machinery for their replication. The hepatitis delta virus of humans has an RNA genome similar to viroids but has protein coat derived from hepatitis B virus and cannot produce one of its own. It is therefore a defective virus and cannot replicate without the help of hepatitis B virus.

The Virophage 'sputnik' infects the Mimivirus and the related Mamavirus, which in turn infect the protozooan *Acanthamoeba castellanii.* These viruses that are dependent on other virus species are called *satellites* and may represent evolutionary intermediates of viroids and viruses. Prions are infectious protein molecules that do not contain DNA or RNA.

They cause an infection in sheep called scrapie and cattle bovine spongiform encephalopathy ("mad cow" disease). In humans they cause kuru and Creutzfeldt-Jakob disease. They are able to replicate because some proteins can exist in two different shapes and the prion changes the normal shape of a host protein into the prion shape. This starts a chain reaction where each prion protein converts many host proteins into more prions, and these new prions then go on to convert even more protein into prions. Although they are fundamentally different from viruses and viroids, their discovery gives credence to the idea that viruses could have evolved from self-replicating molecules.

Computer analysis of viral and host DNA sequences is giving a better understanding of the evolutionary relationships between different viruses and may help identify the ancestors of modern viruses. To date, such analyses have not helped to decide on which of these hypotheses are correct. However, it seems unlikely that all currently known viruses have a common ancestor and viruses have probably arisen numerous times in the past by one or more mechanisms.

Opinions differ on whether viruses are a form of life, or organic structures that interact with living organisms. They have been described as "organisms at the edge of life", since they resemble organisms in that they possess genes and evolve by natural selection, and reproduce by creating multiple copies of themselves through self-assembly. However, although they have genes, they do not have a cellular structure, which is often seen as the basic unit of life. Additionally, viruses do not have their own metabolism, and require a host cell to make new products. They therefore cannot reproduce outside a host cell (although bacterial species such as rickettsia and chlamydia are considered living organisms despite the same limitation). Accepted forms of life use cell division to reproduce, whereas viruses spontaneously assemble within cells, which is analogous to the autonomous growth of crystals. Virus self-assembly within host cells has implications for the study of the origin of life, as it lends further credence to the hypothesis that life could have started as self-assembling organic molecules.

Microbiology

Structure

Viruses display a wide diversity of shapes and sizes, called *morphologies*. Viruses are about 1/100th the size of bacteria. Most viruses that have been studied have a diameter between 10 and 300 nanometres. Some filoviruses have a total length of up to 1400 nm, however their diameters are only about 80 nm. Most viruses are unable to be seen with a light microscope so scanning and transmission electron microscopes are used to visualise virus particles. To increase the contrast between viruses and the background, electron-dense "stains" are used. These are solutions of salts of heavy metals such as tungsten, that scatter the electrons from regions covered with the stain. When virus particles are coated with stain (positive staining), fine detail is obscured. Negative staining overcomes this problem by staining the background only.

A complete virus particle, known as a virion, consists of nucleic acid surrounded by a protective coat of protein called a capsid. These are formed from identical protein subunits called capsomers. Viruses can have a lipid "envelope" derived from the host cell membrane. The capsid is made from proteins encoded by the viral genome and its shape serves as the basis for morphological distinction. Virally coded protein subunits will self-assemble to form a capsid, generally requiring the presence of the virus genome. However, complex viruses code for proteins that assist in the construction of their capsid. Proteins associated with nucleic acid are known as nucleoproteins, and the association of viral capsid proteins with viral nucleic acid is called a nucleocapsid. The capsid and entire virus structure can be mechanically (physically) probed through atomic force microscopy. In general, there are four main morphological virus types:

Helical

These viruses are composed of a single type of capsomer stacked around a central axis to form a helical structure, which may have a central cavity, or hollow tube. This arrangement results in rod-shaped or filamentous virions: these can be short and highly rigid, or long and very flexible. The genetic material, generally single-stranded RNA, but ssDNA in some cases, is bound into the protein helix, by interactions between the negatively charged nucleic acid and positive charges on the protein. Overall, the length of a helical capsid is related to the length of the nucleic acid contained within it and the diameter is dependent on the size and arrangement of capsomers. The well-studied Tobacco mosaic virus is an example of a helical virus.

Icosahedral

Most animal viruses are icosahedral or near-spherical with icosahedral symmetry. A regular icosahedron is the optimum way of forming a closed shell from identical sub-units. The minimum number of identical capsomers required is twelve, each composed of five identical sub-units. Many viruses, such as rotavirus, have more than twelve capsomers and appear spherical but they retain this symmetry. Capsomers at the apices are surrounded by five other capsomers and are called pentons. Capsomers on the triangular faces are surround by six others and are call hexons.

Envelope

Some species of virus envelope themselves in a modified form of one of the cell membranes, either the outer membrane surrounding an infected host cell, or internal membranes such as nuclear membrane or endoplasmic reticulum, thus gaining an outer lipid bilayer known as a viral envelope. This membrane is studded with proteins coded for by the viral genome and host genome; the lipid membrane itself and any carbohydrates present originate entirely from the host. The influenza virus and HIV use this strategy. Most enveloped viruses are dependent on the envelope for their infectivity.

Complex

These viruses possess a capsid that is neither purely helical, nor purely icosahedral, and that may possess extra structures such as protein tails or a complex outer wall. Some bacteriophages, such as Enterobacteria phage T4 have a complex structure consisting of an icosahedral head bound to a helical tail, which may have a hexagonal base plate with protruding protein tail fibres. This tail structure acts like a molecular syringe, attaching to the bacterial host and then injecting the viral genome into the cell.

The poxviruses are large, complex viruses that have an unusual morphology. The viral genome is associated with proteins within a central disk structure known as a nucleoid. The nucleoid is surrounded by a membrane and two lateral bodies of unknown function. The virus has an outer envelope with a thick layer of protein studded over its surface. The whole particle is slightly pleiomorphic, ranging from ovoid to brick shape. Mimivirus is the largest known virus, with a capsid diameter of 400 nm. Protein filaments measuring 100 nm project from the surface. The capsid appears hexagonal under an electron microscope, therefore the capsid is probably icosahedral. Some viruses that infect Archaea have complex structures that are unrelated to any other form of virus, with a wide variety of unusual shapes, ranging from spindle-shaped structures, to viruses that resemble hooked rods, teardrops or even bottles. Other archaeal viruses resemble the tailed bacteriophages, although they can have multiple tail structures.

Genome

An enormous variety of genomic structures can be seen among viral species; as a group they contain more structural genomic diversity than plants, animals, archaea, or bacteria. A virus has either DNA or RNA genes and are called DNA viruses and RNA viruses respectively. By far most viruses have RNA. Plant viruses tend to have single-stranded RNA and bacteriophages tend to have double-stranded DNA. Viral genomes are circular, such as polyomaviruses, or linear, such as adenoviruses. The type of nucleic acid is irrelevant to the shape of the genome. Among RNA viruses, the genome is often divided up into separate parts within the virion and is called *segmented*. Each segment often codes for one protein and they are usually found together in one capsid. Every segment is not required to be in the same virion for the overall virus to be infectious, as demonstrated by the brome mosaic virus.

A viral genome, irrespective of nucleic acid type, is either single-stranded or double-stranded. Single-stranded genomes consist of an unpaired nucleic acid, analogous to one-half of a ladder split down the middle. Double-stranded genomes consist of two complementary paired nucleic acids, analogous to a ladder. Some viruses, such as those belonging to the *Hepadnaviridae*, contain a genome that is partially double-stranded and partially single-stranded.

For viruses with RNA or single-stranded DNA, the strands are said to be either positive-sense (called the plus-strand) or negative-sense (called the minus-strand), depending on whether it is complementary to the viral messenger RNA (mRNA). Positive-sense viral RNA is identical to viral mRNA and thus can be immediately translated by the host cell. Negative-sense viral RNA is complementary to mRNA and thus must be converted to positive-sense RNA by an RNA polymerase before translation. DNA nomenclature is similar to RNA nomenclature, in that the *coding strand* for the viral mRNA is complementary to it ("), and the *non-coding strand* is a copy of it (+).

Genome size varies greatly between species. The smallest viral genomes code for only four proteins and have a mass of about 10^6 Daltons; the largest have a mass of about 10^8 Daltons and code for over one hundred proteins. RNA viruses generally have smaller genome sizes than DNA viruses due to a higher error-rate when replicating, and have a maximum upper size limit. Beyond this limit, errors in the genome when replicating render the virus useless or uncompetitive. To compensate for this, RNA viruses often have segmented genomes where the genome is split into smaller molecules, thus reducing the chance of error. In contrast, DNA viruses generally have larger genomes due to the high fidelity of their replication enzymes.

Viruses undergo genetic change by several mechanisms. These include a process called genetic drift where individual bases in the DNA or RNA mutate to other bases. Most of these point mutations are "silent"—they do not change the protein that the gene encodes—

but others can confer evolutionary advantages such as resistance to antiviral drugs. Antigenic shift is where there is a major change in the genome of the virus. This occurs as a result of recombination or reassortment. When this happens with influenza viruses, pandemics may result. RNA viruses often exist as quasispecies or swarms of viruses of the same species but with slightly different genome nucleoside sequences. Such quasispecies are a prime target for natural selection. Segmented genomes confer evolutionary advantages; different strains of a virus with a segmented genome can shuffle and combine genes and produce progeny viruses or (offspring) that have unique characteristics. This is called reassortment or *viral sex*. Genetic recombination is the process by which a strand of DNA is broken and then joined to the end of a different DNA molecule. This can occur when viruses infect cells simultaneously and studies of viral evolution have shown that recombination has been rampant in the species studied. Recombination is common to both RNA and DNA viruses.

Replication Cycle

Viral populations do not grow through cell division, because they are acellular; instead, they use the machinery and metabolism of a host cell to produce multiple copies of themselves, and they *assemble* in the cell.

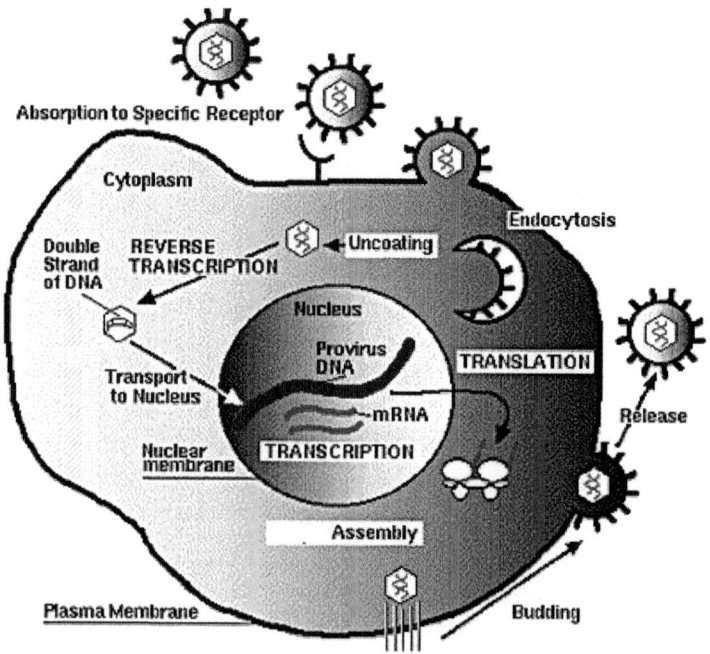

A Typical Virus Replication Cycle

The life cycle of viruses differs greatly between species but there are six *basic* stages in the life cycle of viruses:

- *Attachment* is a specific binding between viral capsid proteins and specific receptors on the host cellular surface. This specificity determines the host range of a virus. For example, HIV infects only human T cells, because its surface protein, gp120, can interact with CD4 and receptors on the T cell's surface. This mechanism has evolved to favour those viruses that only infect cells in which they are capable of replication. Attachment to the receptor can induce the viral-envelope protein to undergo changes that results in the fusion of viral and cellular membranes.

- *Penetration* follows attachment; viruses enter the host cell through receptor mediated endocytosis or membrane fusion. This is often called viral entry. The infection of plant cells is different to that of animal cells. Plants have a rigid cell wall made of cellulose and viruses can only get inside the cells following trauma to the cell wall. Viruses such as tobacco mosaic virus can also move directly in plants, from cell-to-cell, through pores called plasmodesmata. Bacteria, like plants, have strong cell walls that a virus must breach to infect the cell. Some viruses have evolved mechanisms that inject their genome into the bacterial cell while the viral capsid remains outside.

- *Uncoating* is a process in which the viral capsid is degraded by viral enzymes or host enzymes thus releasing the viral genomic nucleic acid.

- *Replication* involves synthesis of viral messenger RNA (mRNA) for viruses except positive sense RNA viruses, viral protein synthesis and assembly of viral proteins and viral genome replication.

- Following the *assembly* of the virus particles, post-translational modification of the viral proteins often occurs. In viruses such as HIV, this modification, (sometimes called maturation), occurs *after* the virus has been released from the host cell.

- Viruses are *released* from the host cell by lysis—a process that kills the cell by bursting its membrane. Enveloped viruses (e.g., HIV) typically are released from the host cell by budding.

During this process the virus acquires its envelope, which is a modified piece of the host's plasma membrane.

The genetic material within viruses, and the method by which the material is replicated, vary between different types of viruses.

DNA Viruses

The genome replication of most DNA viruses takes place in the cell's nucleus. If the cell has the appropriate receptor on its surface, these viruses enter the cell by fusion with the cell membrane or by endocytosis. Most DNA viruses are entirely dependent on the host cell's DNA and RNA synthesising machinery, and RNA processing machinery. The viral genome must cross the cell's nuclear membrane to access this machinery.

RNA Viruses

These viruses are unique because their genetic information is encoded in RNA. Replication usually takes place in the cytoplasm. RNA viruses can be placed into about four different groups depending on their modes of replication. The polarity (whether or not it can be used directly to make proteins) of the RNA largely determines the replicative mechanism, and whether the genetic material is single-stranded or double-stranded. RNA viruses use their own RNA replicase enzymes to create copies of their genomes.

Reverse Transcribing Viruses

These replicate using reverse transcription, which is the formation of DNA from an RNA template. Reverse transcribing viruses containing RNA genomes use a DNA intermediate to replicate, whereas those containing DNA genomes use an RNA intermediate during genome replication. Both types use the reverse transcriptase enzyme to carry out the nucleic acid conversion. Retroviruses often integrate the DNA produced by reverse transcription into the host genome. They are susceptible to antiviral drugs that inhibit the reverse transcriptase enzyme, e.g. zidovudine and lamivudine. An example of the first type is HIV, which is a retrovirus. Examples of the second type are the *Hepadnaviridae*, which includes Hepatitis B virus.

Effects on the Host Cell

The range of structural and biochemical effects that viruses have on the hosts cell is extensive. These are called *cytopathic effects*. Most virus infections eventually result in the death of the host cell. The causes of death include cell lysis, alterations to the cell's surface membrane and apoptosis. Often cell death is caused by cessation of its normal activities due to suppression by virus-specific proteins, not all of which are components of the virus particle.

Some viruses cause no apparent changes to the infected cell. Cells in which the virus is latent and inactive show few signs of infection and often function normally. This causes persistent infections and the virus is often dormant for many months or years. This is often the case with herpes viruses. Viruses, such as Epstein-Barr virus often cause cells to proliferate without causing malignancy, but viruses, such as papillomaviruses are an established cause of cancer.

Classification

Classification seeks to describe the diversity of viruses by naming and grouping them based on similarities. In 1962, Andre Lwoff, Robert Horne, and Paul Tournier were the first to develop a means of virus classification, based on the Linnaean hierarchical system. This system bases classification on phylum, class, order, family, genus, and species. Viruses were grouped according to their shared properties (not of their hosts) and the type of nucleic acid forming their genomes. Later the International Committee on Taxonomy of Viruses was formed.

ICTV Classification

The International Committee on Taxonomy of Viruses (ICTV) developed the current classification system and wrote guidelines that put a greater weight on certain virus properties to maintain family uniformity. A universal system for classifying viruses, and a unified taxonomy, has been established since 1966. The 7th lCTV Report formalised for the first time the concept of the virus species as the lowest taxon (group) in a branching hierarchy of viral taxa. However, at present only a small part of the total diversity of viruses has been studied, with analyses of samples from humans finding that about 20% of the virus sequences recovered have not been seen before, and samples from the environment, such as from seawater and ocean sediments, finding that the large majority of sequences are completely novel.

The general taxonomic structure is as follows:
- Order (-virales)
- Family (-viridae)
- Subfamily (-virinae)

- Genus (*-virus*)
- Species (*-virus*).

In the current (2008) ICTV taxonomy, five orders have been established, the Caudovirales, Herpesvirales, Mononegavirales, Nidovirales, and Picornavirales.

The committee does not formally distinguish between subspecies, strains, and isolates. In total there are 5 orders, 82 families, 11 subfamilies, 307 genera, 2,083 species and about 3,000 types yet unclassified.

Baltimore Classification

The Nobel Prize-winning biologist David Baltimore devised the Baltimore classification system. The ICTV classification system is used in conjunction with the Baltimore classification system in modern virus classification.

The Baltimore classification of viruses is based on the mechanism of mRNA production. Viruses must generate mRNAs from their genomes to produce proteins and replicate themselves, but different mechanisms are used to achieve this in each virus family.

Viral genomes may be single-stranded (ss) or double-stranded (ds), RNA or DNA, and may or may not use reverse transcriptase (RT). Additionally, ssRNA viruses may be either sense (+) or antisense (-). This classification places viruses into seven groups:

- I: dsDNA viruses (e.g. Adenoviruses, Herpesviruses, Poxviruses)
- II: ssDNA viruses (+)sense DNA (e.g. Parvoviruses)
- III: dsRNA viruses (e.g. Reoviruses)
- IV: (+)ssRNA viruses (+)sense RNA (e.g. Picornaviruses, Togaviruses)
- V: (-)ssRNA viruses (-)sense RNA (e.g. Orthomyxoviruses, Rhabdoviruses)
- VI: ssRNA-RT viruses (+)sense RNA with DNA intermediate in life-cycle (e.g. Retroviruses)
- VII: dsDNA-RT viruses (e.g. Hepadnaviruses).

As an example of viral classification, the chicken pox virus, varicella zoster (VZV), belongs to the order Herpesvirales, family *Herpesviridae*, subfamily *Alphaherpesvirinae*, and genus *Varicellovirus*. VZV is in Group I of the Baltimore Classification because it is a dsDNA virus that does not use reverse transcriptase.

Viruses and Human Disease

Examples of common human diseases caused by viruses include the common cold, influenza, chickenpox and cold sores. Many serious diseases such as ebola, AIDS, avian influenza and SARS are caused by viruses. The relative ability of viruses to cause disease is described in terms of virulence. Other diseases are under investigation as to whether they too have a virus as the causative agent, such as the possible connection between human herpes virus six (HHV6) and neurological diseases such as multiple sclerosis and chronic fatigue syndrome. There is current controversy over whether the borna virus, previously thought to cause neurological diseases in horses, could be responsible for psychiatric illnesses in humans.

Viruses have different mechanisms by which they produce disease in an organism, which largely depends on the viral species. Mechanisms at the cellular level primarily include cell lysis, the breaking open and subsequent death of the cell. In multicellular organisms, if enough cells die the whole organism will start to suffer the effects. Although viruses cause disruption of healthy homeostasis, resulting in disease, they may exist relatively harmlessly within an organism. An example would include the ability of the herpes simplex virus, which cause cold sores, to remain in a dormant state within the human body. This is called latency and is a characteristic of the all herpes viruses including the Epstein-Barr virus, which causes glandular fever, and the varicella zoster virus, which causes chicken pox.

The large majority of people have been infected with at least one of these types of herpes virus. However, these latent viruses might sometimes be beneficial, as the presence of the virus can increase immunity against bacterial pathogens, such as *Yersinia pestis*. On the other hand, latent chickenpox infections return in later life as the disease called shingles.

Some viruses can cause life-long or chronic infections, where the viruses continue to replicate in the body despite the hosts' defence mechanisms. This is common in hepatitis B virus and hepatitis C virus infections. People chronically infected are known as carriers, as they serve as reservoirs of infectious virus. In populations with a high proportion of carriers, the disease is said to be endemic. In contrast to acute lytic viral infections this persistence implies compatible interactions with the host organism. Persistent viruses may even broaden the evolutionary potential of host species.

Epidemiology

Viral epidemiology is the branch of medical science that deals with the transmission and control of virus infections in humans. Transmission of viruses can be vertical, that is from mother to child, or horizontal, which means from person to person.

Examples of vertical transmission include hepatitis B virus and HIV where the baby is born already infected with the virus. Another, more rare, example is the varicella zoster virus, which although causing relatively mild infections in humans, can be fatal to the foetus and newly born baby. Horizontal transmission is the most common mechanism of spread of viruses in populations.

Transmission can be exchange of blood by sexual activity, e.g. HIV, hepatitis B and hepatitis C; by mouth by exchange of saliva, e.g. Epstein-Barr virus, or from contaminated food or water, e.g. norovirus; by breathing in viruses in the form of aerosols, e.g. influenza virus; and by insect vectors such as mosquitoes, e.g. dengue. The rate or speed of transmission of viral infections depends on factors that include population density, the number of susceptible individuals, (i.e. those who are not immune), the quality of health care and the weather.

Epidemiology is used to break the chain of infection in populations during outbreaks of viral diseases. Control measures are used that are based on knowledge of how the virus is transmitted. It is important to find the source, or sources, of the outbreak and to identify the virus. Once the virus has been identified, the chain of transmission can sometimes be broken by vaccines. When vaccines are not available sanitation and disinfection can be effective. Often infected people are isolated from the rest of the community and those that have been exposed to the virus placed in quarantine.

To control the outbreak of foot and mouth disease in cattle in Britain in 2001, thousands of cattle were slaughtered. Most viral infections of humans and other animals have incubation periods during which the infection causes no signs or symptoms. Incubation periods for viral diseases range from a few days to weeks but are known for most infections. Somewhat overlapping, but mainly following the incubation period, there is a *period of communicability*; a time when an infected individual or animal is contagious and can infect another person or animal. This too is known for many viral infections and knowledge the length of both periods is important in the control of outbreaks. When outbreaks cause an unusually high proportion of cases in a population, community or region they are called epidemics. If outbreaks spread worldwide they are called pandemics.

Epidemics and Pandemics

Native American populations were devastated by contagious diseases, particularly smallpox, brought to the Americas by European colonists. It is unclear how many Native Americans were killed by foreign diseases after the arrival of Columbus in the Americas, but the numbers have been estimated to be close to 70% of the indigenous population. The damage done by this disease significantly aided European attempts to displace and conquer the native population.

A pandemic is a worldwide epidemic. The 1918 flu pandemic, commonly referred to as the Spanish flu, was a category 5 influenza pandemic caused by an unusually severe and deadly influenza A virus. The victims were often healthy young adults, in contrast to most influenza outbreaks, which predominantly affect juvenile, elderly, or otherwise weakened patients.

The Spanish flu pandemic lasted from 1918 to 1919. Older estimates say it killed 40–50 million people, while more recent research suggests that it may have killed as many as 100 million people, or 5% of the world's population in 1918. Most researchers believe that HIV originated in sub-Saharan Africa during the twentieth century; it is now a pandemic, with an estimated 38.6 million people now living with the disease worldwide. The Joint United Nations Programme on HIV/AIDS (UNAIDS) and the World Health Organization (WHO) estimate that AIDS has killed more than 25 million people since it was first recognised on June 5, 1981, making it one of the most destructive epidemics in recorded history. In 2007 there were 2.7 million new HIV infections and 2 million HIV-related deaths.

Several highly lethal viral pathogens are members of the *Filoviridae*. Filoviruses are filament-like viruses that cause viral hemorrhagic fever, and include the ebola and marburg viruses. The Marburg virus attracted widespread press attention in April 2005 for an outbreak in Angola. Beginning in October 2004 and continuing into 2005, the outbreak was the world's worst epidemic of any kind of viral hemorrhagic fever.

Cancer

Viruses are an established cause of cancer in humans and other species. Viral cancers only occur in a minority of infected persons (or animals). Cancer viruses come from a range of virus families, including both RNA and DNA viruses, and so there is no single type of "oncovirus" (an obsolete term originally used for acutely-transforming retroviruses). The development of cancer is determined by a variety of factors such as host immunity and mutations in the host.

Viruses accepted to cause human cancers inclue some genotypes of human papillomavirus, hepatitis B virus, hepatitis C virus, Epstein-Barr virus, Kaposi's sarcoma-associated herpesvirus and human T-lymphotropic virus. The most recently discovered human cancer virus is a polyomavirus (Merkel cell polyomavirus) that causes most cases of a rare form of skin cancer called Merkel cell carcinoma. Hepatitis viruses can develop into a chronic viral infection that leads to liver cancer. Infection by human T-lymphotropic virus can lead to tropical spastic paraparesis and adult T-cell leukemia. Human papillomaviruses are an established cause of cancers of cervix, skin, anus, and penis. Within the *Herpesviridae*, Kaposi's sarcoma-associated herpesvirus causes Kaposi's sarcoma and body cavity lymphoma, and Epstein–Barr virus causes Burkitt's lymphoma, Hodgkin's lymphoma, B lymphoproliferative disorder and nasopharyngeal carcinoma. Merkel cell polyomavirus closely related to SV40 and mouse polyomaviruses that have been used as animal models for cancer viruses for over 50 years.

Host Defence Mechanisms

The body's first line of defence against viruses is the innate immune system. This comprises cells and other mechanisms that defend the host from infection in a non-specific manner. This means that the cells of the innate system recognise, and respond to, pathogens in a generic way, but unlike the adaptive immune system, it does not confer long-lasting or protective immunity to the host.

RNA interference is an important innate defence against viruses. Many viruses have a replication strategy that involves double-stranded RNA (dsRNA). When such a virus infects a cell, it releases its RNA molecule or molecules, which immediately bind to a protein complex called dicer that cuts the RNA into smaller pieces. A biochemical pathway called the RISC complex is activated, which degrades the viral mRNA and the cell survives the infection. Rotaviruses avoid this mechanism by not uncoating fully inside the cell and by releasing newly produced mRNA through pores in the particle's inner capsid. The genomic dsRNA remains protected inside the core of the virion.

When the adaptive immune system of a vertebrate encounters a virus, it produces specific antibodies that bind to the virus and render it non-infectious. This is called humoral immunity. Two types of antibodies are important. The first called IgM is highly effective at neutralizing viruses but is only produced by the cells of the immune system for a few weeks. The second, called, IgG is produced indefinitely. The presence of IgM in the blood of the host is used to test for acute infection, whereas IgG indicates an infection sometime in the past. IgG antibody is measured when tests for immunity are carried out.

A second defence of vertebrates against viruses is called cell-mediated immunity and involves immune cells known as T cells. The body's cells constantly display short fragments of their proteins on the cell's surface, and if a T cell recognises a suspicious viral fragment there, the host cell is destroyed by *T killer* cells and the virus-specific T-cells proliferate. Cells such as the macrophage are specialists at this antigen presentation. The production of interferon is an important host defence mechanism. This is a hormone produced by the body when viruses are present. Its role in immunity is complex, but it eventually stops the viruses from reproducing by killing the infected cell and its close neighbours

Not all virus infections produce a protective immune response in this way. HIV evades the immune system by constantly changing the amino acid sequence of the proteins on the surface of the virion. These persistent viruses evade immune control by sequestration, blockade of antigen presentation, cytokine resistance, evasion of natural killer cell activities, escape from apoptosis, and antigenic shift. Other viruses, called *neurotropic viruses*, are disseminated by neural spread where the immune system may be unable to reach them.

Prevention and Treatment

Because viruses use the machinery of a host cell to reproduce and reside within them, they are difficult to eliminate without killing the host cell. The most effective medical approaches to viral diseases so far are vaccinations to provide resistance to infection, and antiviral drugs.

Vaccines

Vaccination is a cheap and effective way of preventing infections by viruses. Vaccines were used to prevent viral infections long before the discovery of the actual viruses. Their use has resulted in a dramatic decline in morbidity (illness) and mortality (death) associated with viral infections such as polio, measles, mumps and rubella. Smallpox infections have been eradicated. Currently vaccines are available to prevent over thirteen viral infections of humans, and more are used to prevent viral infections of animals. Vaccines can consist of live-attenuated or killed viruses, or viral proteins (antigens). Live vaccines contain weakened forms of the virus that causes the disease. Such viruses are called attenuated. Live vaccines can be dangerous when given to people with a weak immunity, (who are described as immunocompromised), because in these people, the weakened virus can cause the original disease. Biotechnology and genetic engineering techniques are used to produce subunit vaccines. These vaccines use only the capsid proteins of the virus. Hepatitis B vaccine is an example of this type of vaccine. Subunit vaccines are safe for immunocompromised patients because they cannot cause the disease. However, the yellow fever virus vaccine, a live-attenuated strain called 17D, is probably the safest and most effective vaccine ever generated.

Antiviral Drugs

Over the past twenty years, the development of antiviral drugs has increased rapidly. This has been driven by the AIDS pandemic. Antiviral drugs are often nucleoside analogues, (fake DNA building blocks), which viruses incorporate into their genomes during replication. The life-cycle of the virus is then halted because the newly synthesised DNA is inactive. This is because these analogues lack the hydroxyl groups, which, along with phosphorus atoms, link together to form the strong "backbone" of the DNA molecule.

This is called DNA chain termination. Examples of nucleoside analogues are aciclovir for Herpes simplex virus infections and lamivudine for HIV and Hepatitis B virus infections. Aciclovir is one of the oldest and most frequently prescribed antiviral drugs. Other antiviral drugs in use target different stages of the viral life cycle. HIV is dependent on a proteolytic enzyme called the HIV-1 protease for it to become fully infectious. There is a large class of drugs called protease inhibitors that inactivate this enzyme.

Hepatitis C is caused by an RNA virus. In 80% of people infected, the disease is chronic, and without treatment, they are infected for the remainder of their lives. However, there is now an effective treatment that uses the nucleoside analogue drug ribavirin combined with interferon. The treatment of chronic carriers of the hepatitis B virus by using a similar strategy using lamivudine has been developed.

Infection in other Species

Viruses infect all cellular life and, although viruses occur universally, each cellular species has its own specific range that often infect only that species. Viruses are important pathogens of livestock. Diseases such as Foot and Mouth Disease and bluetongue are caused by viruses. Companion animals such as cats, dogs, and horses, if not vaccinated, are susceptible to serious viral infections. Canine parvovirus is caused by a small DNA virus and infections are often fatal in pups. Like all invertebrates, the honey bee is susceptible to many viral infections. Fortunately, most viruses co-exist harmlessly in their host and cause no signs or symptoms of disease.

Plants

There are many types of plant virus, but often they cause only a loss of yield, and it is not economically viable to try to control them. Plant viruses are often spread from plant to plant by organisms, known as *vectors*. These are normally insects, but some fungi, nematode worms and single-celled organisms have been shown to be vectors. When control of plant virus infections is considered economical, for perennial fruits for example, efforts are concentrated on killing the vectors and removing alternate hosts such as weeds. Plant viruses are harmless to humans and other animals because they can reproduce only in living plant cells.

Plants have elaborate and effective defence mechanisms against viruses. One of the most effective is the presence of so-called resistance (R) genes. Each R gene confers resistance to a particular virus by triggering localised areas of cell death around the infected cell, which can often be seen with the unaided eye as large spots. This stops the infection from spreading. RNA interference is also an effective defence in plants. When they are infected, plants often produce natural disinfectants that kill viruses, such as salicylic acid, nitric oxide, and reactive oxygen molecules.

Bacteria

Bacteriophages are an extremely common and diverse group of viruses. For example, bacteriophages are the most common form of biological entity in aquatic environments, with up to ten times more of these viruses in the oceans than bacteria, reaching levels of 250,000,000 bacteriophages per millilitre of seawater. These viruses infect specific bacteria by binding to surface receptor molecules and then entering the cell. Within a short amount of time, in some cases just minutes, bacterial polymerase starts translating viral mRNA into protein. These proteins go on to become either new virions within the cell, helper proteins, which help assembly of new virions, or proteins involved in cell lysis. Viral enzymes aid in the breakdown of the cell membrane, and, in the case of the T4 phage, in just over twenty minutes after injection over three hundred phages could be released.

The major way bacteria defend themselves from bacteriophages is by producing enzymes that destroy foreign DNA. These enzymes, called restriction endonucleases, cut up the viral DNA that bacteriophages inject into bacterial cells. Bacteria also contain a system that uses CRISPR sequences to retain fragments of the genomes of viruses that the bacteria have come into contact with in the past, which allows them to block the virus's replication through a form of RNA interference. This genetic system provides bacteria with acquired immunity to infection.

Archaea

Some viruses replicate within archaea: these are double-stranded DNA viruses with unusual and sometimes unique shapes. These viruses have been studied in most detail in the thermophilic archaea, particularly the orders Sulfolobales and Thermoproteales. Defences against these viruses may involve RNA interference from repetitive DNA sequences within archaean genomes that are related to the genes of the viruses.

Applications

Life Sciences and Medicine

Viruses are important to the study of molecular and cellular biology as they provide simple systems that can be used to manipulate and investigate the functions of cells. The study and use of viruses have provided valuable information about aspects of cell biology. For example, viruses have been useful in the study of genetics and helped our understanding of the basic mechanisms of molecular genetics, such as DNA replication, transcription, RNA processing, translation, protein transport, and immunology.

Geneticists often use viruses as vectors to introduce genes into cells that they are studying. This is useful for making the cell produce a foreign substance, or to study the effect of introducing a new gene into the genome. In similar fashion, virotherapy uses viruses as vectors to treat various diseases, as they can specifically target cells and

DNA. It shows promising use in the treatment of cancer and in gene therapy. Eastern European scientists have used phage therapy as an alternative to antibiotics for some time, and interest in this approach is increasing, due to the high level of antibiotic resistance now found in some pathogenic bacteria.

Materials Science and Nanotechnology

Current trends in nanotechnology promise to make much more versatile use of viruses. From the viewpoint of a materials scientist, viruses can be regarded as organic nanoparticles. Their surface carries specific tools designed to cross the barriers of their host cells. The size and shape of viruses, and the number and nature of the functional groups on their surface, is precisely defined. As such, viruses are commonly used in materials science as scaffolds for covalently linked surface modifications. A particular quality of viruses is that they can be tailored by directed evolution. The powerful techniques developed by life sciences are becoming the basis of engineering approaches towards nanomaterials, opening a wide range of applications far beyond biology and medicine.

Because of their size, shape, and well-defined chemical structures, viruses have been used as templates for organizing materials on the nanoscale. Recent examples include work at the Naval Research Laboratory in Washington, DC, using Cowpea Mosaic Virus (CPMV) particles to amplify signals in DNA microarray based sensors. In this application, the virus particles separate the fluorescent dyes used for signaling in order to prevent the formation of non-fluorescent dimers that act as quenchers. Another example is the use of CPMV as a nanoscale breadboard for molecular electronics.

Weapons

The ability of viruses to cause devastating epidemics in human societies has led to the concern that viruses could be weaponised for biological warfare. Further concern was raised by the successful recreation of the infamous 1918 influenza virus in a laboratory. The smallpox virus devastated numerous societies throughout history before its eradication. There are officially only two centres in the world which keep stocks of smallpox virus - the Russian Vector laboratory, and the United States Centres for Disease Control. But fears that it may be used as a weapon are not totally unfounded; the vaccine for smallpox is not safe — during the years before the eradication of smallpox disease more people became seriously ill as a result of vaccination than did people from smallpox — and smallpox vaccination is no longer universally practiced. Thus, much of the modern human population has almost no established resistance to smallpox.

Bacteria

The bacteria are a large group of unicellular microorganisms. Typically a few micrometres in length, bacteria have a wide range of shapes, ranging from spheres to rods and spirals. Bacteria are ubiquitous in every habitat on Earth, growing in soil, acidic hot springs, radioactive waste, water, and deep in the Earth's crust, as well as in organic matter and the live bodies of plants and animals. There are typically 40 million bacterial cells in a gram of soil and a million bacterial cells in a millilitre of fresh water; in all, there are approximately five nonillion (5×10^{30}) bacteria on Earth, forming much of the world's biomass. Bacteria are vital in recycling nutrients, with many steps in nutrient cycles depending on these organisms, such as the fixation of nitrogen from the atmosphere and putrefaction. However, most bacteria have not been characterized, and only about half of the phyla of bacteria have species that can be grown in the laboratory. The study of bacteria is known as bacteriology, a branch of microbiology.

There are approximately ten times as many bacterial cells in the human flora of bacteria as there are human cells in the body, with large numbers of bacteria on the skin and as gut flora. The vast majority of the bacteria in the body are rendered harmless by the protective effects of the immune system, and a few are beneficial. However, a few species of bacteria are pathogenic and cause infectious diseases, including cholera, syphilis, anthrax, leprosy and bubonic plague. The most common fatal bacterial diseases are respiratory infections, with tuberculosis alone killing about 2 million people a year, mostly in sub-Saharan Africa. In developed countries, antibiotics are used to treat bacterial infections and in agriculture, so antibiotic resistance is becoming common. In industry, bacteria are important in sewage treatment, the production of cheese and yoghurt through fermentation, as well as in biotechnology, and the manufacture of antibiotics and other chemicals.

Once regarded as plants constituting the class Schizomycetes, bacteria are now classified as prokaryotes. Unlike cells of animals and other eukaryotes, bacterial cells do not contain a nucleus and rarely harbour membrane-bound organelles. Although the term *bacteria* traditionally included all prokaryotes, the scientific classification changed after the discovery in the 1990s that prokaryotes consist of two very different groups of organisms that evolved independently from an ancient common ancestor. These evolutionary domains are called Bacteria and Archaea.

History of Bacteriology

Bacteria were first observed by Antonie van Leeuwenhoek in 1676, using a single-lens microscope of his own design. He called them "animalcules" and published his observations in a series of letters to the Royal Society. The name *bacterium* was introduced much later, by Christian Gottfried Ehrenberg in 1838.

Louis Pasteur demonstrated in 1859 that the fermentation process is caused by the growth of microorganisms, and that this growth is not due to spontaneous generation. (Yeasts and molds, commonly associated with fermentation, are not bacteria, but rather fungi.)

Along with his contemporary, Robert Koch, Pasteur was an early advocate of the germ theory of disease. Robert Koch was a pioneer in medical microbiology and worked on cholera, anthrax and tuberculosis. In his research into tuberculosis, Koch finally proved the germ theory, for which he was awarded a Nobel Prize in 1905. In *Koch's postulates*, he set out criteria to test if an organism is the cause of a disease; these postulates are still used today.

Though it was known in the nineteenth century that bacteria are the cause of many diseases, no effective antibacterial treatments were available. In 1910, Paul Ehrlich developed the first antibiotic, by changing dyes that selectively stained *Treponema pallidum*—the spirochaete that causes syphilis—into compounds that selectively killed the pathogen. Ehrlich had been awarded a 1908 Nobel Prize for his work on immunology, and pioneered the use of stains to detect and identify bacteria, with his work being the basis of the Gram stain and the Ziehl-Neelsen stain.

A major step forward in the study of bacteria was the recognition in 1977 by Carl Woese that archaea have a separate line of evolutionary descent from bacteria. This new phylogenetic taxonomy was based on the sequencing of 16S ribosomal RNA, and divided prokaryotes into two evolutionary domains, as part of the three-domain system.

Origin and Early Evolution

The ancestors of modern bacteria were single-celled microorganisms that were the first forms of life to develop on earth, about 4 billion years ago. For about 3 billion years, all organisms were microscopic, and bacteria and archaea were the dominant forms of life. Although bacterial fossils exist, such as stromatolites, their lack of distinctive morphology prevents them from being used to examine the history of bacterial evolution, or to date the time of origin of a particular bacterial species. However, gene sequences can be used to reconstruct the bacterial phylogeny, and these studies indicate that bacteria diverged first from the archaeal/eukaryotic lineage. The most recent common ancestor of bacteria and archaea was probably a hyperthermophile that lived about 2.5 billion–3.2 billion years ago.

Bacteria were also involved in the second great evolutionary divergence, that of the archaea and eukaryotes. Here, eukaryotes resulted from ancient bacteria entering into endosymbiotic associations with the ancestors of eukaryotic cells, which were themselves possibly related to the Archaea. This involved the engulfment by proto-eukaryotic cells of alpha-proteobacterial symbionts to form either mitochondria or hydrogenosomes, which are still being found in all known Eukarya (sometimes in highly reduced form, e.g. in ancient "amitochondrial" protozoa). Later on, some eukaryotes that already contained mitochondria also engulfed cyanobacterial-like organisms. This led to the formation of chloroplasts in algae and plants. There are also some algae that originated from even later endosymbiotic events. Here, eukaryotes engulfed a eukaryotic algae that developed into a "second-generation" plastid. This is known as secondary endosymbiosis.

Morphology

Bacteria display a wide diversity of shapes and sizes, called *morphologies*. Bacterial cells are about one tenth the size of eukaryotic cells and are typically 0.5–5.0 micrometres in length. However, a few species–for example *Thiomargarita namibiensis* and *Epulopiscium fishelsoni*–are up to half a millimetre long and are visible to the unaided eye. Among the smallest bacteria are members of the genus *Mycoplasma*, which measure only 0.3 micrometres, as small as the largest viruses. Some bacteria may be even smaller, but these ultramicrobacteria are not well-studied.

Most bacterial species are either spherical, called cocci (*sing.* coccus, from Greek *kókkos*, grain, seed) or rod-shaped, called bacilli (*sing.* bacillus, from Latin *baculus*, stick). Some rod-shaped bacteria, called vibrio, are slightly curved or comma-shaped; others, can be spiral-shaped, called spirilla, or tightly coiled, called spirochaetes. A small number of species even have tetrahedral or cuboidal shapes. More recently, bacteria were discovered deep under the Earth's crust that grow as long rods with a star-shaped cross-section. The large surface area to volume ratio of this morphology may give these bacteria an advantage in nutrient-poor environments. This wide variety of shapes is determined by the bacterial cell wall and cytoskeleton, and is important because it can influence the ability of bacteria to acquire nutrients, attach to surfaces, swim through liquids and escape predators.

Many bacterial species exist simply as single cells, others associate in characteristic patterns: *Neisseria* form diploids (pairs), *Streptococcus* form chains, and *Staphylococcus* group together in "bunch of grapes" clusters. Bacteria can also be elongated to form filaments, for example the Actinobacteria. Filamentous bacteria are often surrounded by a sheath that contains many individual cells. Certain types, such as species of the genus *Nocardia*, even form complex, branched filaments, similar in appearance to fungal mycelia.

Bacteria often attach to surfaces and form dense aggregations called biofilms or bacterial mats. These films can range from a few micrometers in thickness to up to half a meter in depth, and may contain multiple species of bacteria, protists and archaea. Bacteria living in biofilms display a complex arrangement of cells and extracellular components, forming secondary structures such as microcolonies, through which there are networks of channels to enable better diffusion of nutrients. In natural environments, such as soil or the surfaces of plants, the majority of bacteria are bound to surfaces in biofilms. Biofilms are also important in medicine, as these structures are often present during chronic bacterial infections or in infections of implanted medical devices, and bacteria protected within biofilms are much harder to kill than individual isolated bacteria.

Even more complex morphological changes are sometimes possible. For example, when starved of amino acids, Myxobacteria detect surrounding cells in a process known as quorum sensing, migrate towards each other, and aggregate to form fruiting bodies up to 500 micrometres long and containing approximately 100,000 bacterial cells. In these fruiting bodies, the bacteria perform separate tasks; this type of cooperation is a simple type of multicellular organisation. For example, about one in 10 cells migrate to the top of these fruiting bodies and differentiate into a specialised dormant state called myxospores, which are more resistant to drying and other adverse environmental conditions than are ordinary cells.

Cellular Structure

Intracellular Structures

The bacterial cell is surrounded by a lipid membrane, or cell membrane, which encloses the contents of the cell and acts as a barrier to hold nutrients, proteins and other essential components of the cytoplasm within the cell. As they are prokaryotes, bacteria do not tend to have membrane-bound organelles in their cytoplasm and thus contain few large intracellular structures. They consequently lack a nucleus, mitochondria, chloroplasts and the other organelles present in eukaryotic cells, such as the Golgi apparatus and endoplasmic reticulum. Bacteria were once seen as simple bags of cytoplasm, but elements such as prokaryotic cytoskeleton, and the localization of proteins to specific locations within the cytoplasm have been found to show levels of complexity. These subcellular compartments have been called "bacterial hyperstructures".

Micro-compartments such as carboxysome provides a further level of organization, which are compartments within bacteria that are surrounded by polyhedral protein shells, rather than by lipid membranes. These "polyhedral organelles" localize and compartmentalize bacterial metabolism, a function performed by the membrane-bound organelles in eukaryotes.

Many important biochemical reactions, such as energy generation, occur by concentration gradients across membranes, a potential difference also found in a battery. The general lack of internal membranes in bacteria means reactions such as electron transport occur across the cell membrane between the cytoplasm and the periplasmic space. However, in many photosynthetic bacteria the plasma membrane is highly folded and fills most of the cell with layers of light-gathering membrane. These light-gathering complexs may even form lipid-enclosed structures called chlorosomes in green sulfur bacteria. Other proteins import nutrients across the cell membrane, or to expel undesired molecules from the cytoplasm.

Bacteria do not have a membrane-bound nucleus, and their genetic material is typically a single circular chromosome located in the cytoplasm in an irregularly shaped body called the nucleoid. The nucleoid contains the chromosome with associated proteins and RNA.

The order Planctomycetes are an exception to the general absence of internal membranes in bacteria, because they have a membrane around their nucleoid and contain other membrane-bound cellular structures. Like all living organisms, bacteria contain ribosomes for the production of proteins, but the structure of the bacterial ribosome is different from those of eukaryotes and Archaea.

Some bacteria produce intracellular nutrient storage granules, such as glycogen, polyphosphate, sulfur or polyhydroxyalkanoates. These granules enable bacteria to store compounds for later use. Certain bacterial species, such as the photosynthetic Cyanobacteria, produce internal gas vesicles, which they use to regulate their buoyancy - allowing them to move up or down into water layers with different light intensities and nutrient levels.

Extracellular Structures

Around the outside of the cell membrane is the bacterial cell wall. Bacterial cell walls are made of peptidoglycan (called murein in older sources), which is made from polysaccharide chains cross-linked by unusual peptides containing D-amino acids. Bacterial cell walls are different from the cell walls of plants and fungi, which are made of cellulose and chitin, respectively. The cell wall of bacteria is also distinct from that of Archaea, which do not contain peptidoglycan. The cell wall is essential to the survival of many bacteria, and the antibiotic penicillin is able to kill bacteria by inhibiting a step in the synthesis of peptidoglycan.

There are broadly speaking two different types of cell wall in bacteria, called Gram-positive and Gram-negative. The names originate from the reaction of cells to the Gram stain, a test long-employed for the classification of bacterial species.

Gram-positive bacteria possess a thick cell wall containing many layers of peptidoglycan and teichoic acids. In contrast, Gram-negative bacteria have a relatively thin cell wall consisting of a few layers of peptidoglycan surrounded by a second lipid membrane containing lipopolysaccharides and lipoproteins. Most bacteria have the Gram-negative cell wall, and only the Firmicutes and Actinobacteria (previously known as the low G+C and high G+C Gram-positive bacteria, respectively) have the alternative Gram-positive arrangement. These differences in structure can produce differences in antibiotic susceptibility; for instance, vancomycin can kill only Gram-positive bacteria and is ineffective against Gram-negative pathogens, such as

Haemophilus influenzae or *Pseudomonas aeruginosa*. In many bacteria an S-layer of rigidly arrayed protein molecules covers the outside of the cell. This layer provides chemical and physical protection for the cell surface and can act as a macromolecular diffusion barrier. S-layers have diverse but mostly poorly understood functions, but are known to act as virulence factors in *Campylobacter* and contain surface enzymes in *Bacillus stearothermophilus*.

Flagella are rigid protein structures, about 20 nanometres in diameter and up to 20 micrometres in length, that are used for motility. Flagella are driven by the energy released by the transfer of ions down an electrochemical gradient across the cell membrane.

Fimbriae are fine filaments of protein, just 2–10 nanometres in diameter and up to several micrometers in length. They are distributed over the surface of the cell, and resemble fine hairs when seen under the electron microscope. Fimbriae are believed to be involved in attachment to solid surfaces or to other cells and are essential for the virulence of some bacterial pathogens. Pili (*sing.* pilus) are cellular appendages, slightly larger than fimbriae, that can transfer genetic material between bacterial cells in a process called conjugation.

Capsules or slime layers are produced by many bacteria to surround their cells, and vary in structural complexity: ranging from a disorganised slime layer of extra-cellular polymer, to a highly structured capsule or glycocalyx. These structures can protect cells from engulfment by eukaryotic cells, such as macrophages. They can also act as antigens and be involved in cell recognition, as well as aiding attachment to surfaces and the formation of biofilms.

The assembly of these extracellular structures is dependent on bacterial secretion systems. These transfer proteins from the cytoplasm into the periplasm or into the environment around the cell. Many types of secretion systems are known and these structures are often essential for the virulence of pathogens, so are intensively studied.

Endospores

Certain genera of Gram-positive bacteria, such as *Bacillus*, *Clostridium*, *Sporohalobacter*, *Anaerobacter* and *Heliobacterium*, can form highly resistant, dormant structures called endospores. In almost all cases, one endospore is formed and this is not a reproductive process, although *Anaerobacter* can make up to seven endospores in a single cell. Endospores have a central core of cytoplasm containing DNA and ribosomes surrounded by a cortex layer and protected by an impermeable and rigid coat. Endospores show no detectable metabolism and can survive extreme physical and chemical stresses, such as high levels of UV light, gamma radiation, detergents, disinfectants, heat, pressure and desiccation. In this dormant state, these organisms may remain viable for millions of years, and endospores even allow bacteria to survive exposure to the vacuum and radiation in space. Endospore-forming bacteria can also cause disease: for example, anthrax can be contracted by the inhalation of *Bacillus anthracis* endospores, and contamination of deep puncture wounds with *Clostridium tetani* endospores causes tetanus.

Metabolism

Further Information: Microbial Metabolism

Bacteria exhibit an extremely wide variety of metabolic types. The distribution of metabolic traits within a group of bacteria has traditionally been used to define their taxonomy, but these traits often do not correspond with modern genetic classifications. Bacterial metabolism is classified into nutritional groups on the basis of three major criteria: the kind of energy used for growth, the source of carbon, and the electron donors used for growth. An additional criterion of respiratory microorganisms are the electron acceptors used for aerobic or anaerobic respiration.

Carbon metabolism in bacteria is either heterotrophic, where organic carbon compounds are used as carbon sources, or autotrophic, meaning that cellular carbon is obtained by fixing carbon dioxide. Heterotrophic bacteria include parasitic types. Typical autotrophic bacteria are phototrophic cyanobacteria, green sulfur-bacteria and some purple bacteria, but also many chemolithotrophic species, such as nitrifying or sulfur-oxidising bacteria. Energy metabolism of bacteria is either based on phototrophy, the use of light through photosynthesis, or on chemotrophy, the use of chemical substances for energy, which are mostly oxidised at the expense of oxygen or alternative electron acceptors (aerobic/anaerobic respiration).

Finally, bacteria are further divided into lithotrophs that use inorganic electron donors and organotrophs that use organic compounds as electron donors. Chemotrophic organisms use the respective electron donors for energy conservation (by aerobic/anaerobic respiration or fermentation) and biosynthetic reactions (e.g. carbon dioxide fixation), whereas phototrophic organisms use them only for biosynthetic purposes. Respiratory organisms use chemical compounds as a source of energy by taking electrons from the reduced substrate and transferring them to a terminal electron acceptor in a redox reaction. This reaction releases energy that can be used to synthesise ATP and drive metabolism. In aerobic organisms, oxygen is used as the electron acceptor. In anaerobic organisms other inorganic compounds, such as nitrate, sulfate or carbon dioxide are used as electron acceptors. This leads to the ecologically important processes of denitrification, sulfate reduction and acetogenesis, respectively.

Another way of life of chemotrophs in the absence of possible electron acceptors is fermentation, where the electrons taken from the reduced substrates are transferred to oxidised intermediates to generate reduced fermentation products (e.g. lactate, ethanol, hydrogen, butyric acid). Fermentation is possible, because the energy content of the substrates is higher than that of the products, which allows the organisms to synthesise ATP and drive their metabolism.

These processes are also important in biological responses to pollution; for example, sulfate-reducing bacteria are largely responsible for the production of the highly toxic forms of mercury (methyl- and dimethylmercury) in the environment. Non-respiratory anaerobes use fermentation to generate energy and reducing power, secreting metabolic by-products (such as ethanol in brewing) as waste. Facultative anaerobes can switch between fermentation and different terminal electron acceptors depending on the environmental conditions in which they find themselves.

Lithotrophic bacteria can use inorganic compounds as a source of energy. Common inorganic electron donors are hydrogen, carbon monoxide, ammonia (leading to nitrification), ferrous iron and other reduced metal ions, and several reduced sulfur compounds. Unusually, the gas methane can be used by methanotrophic bacteria as both a source of electrons and a substrate for carbon anabolism. In both aerobic phototrophy and chemolithotrophy, oxygen is used as a terminal electron acceptor, while under anaerobic conditions inorganic compounds are used instead. Most lithotrophic organisms are autotrophic, whereas organotrophic organisms are heterotrophic. In addition to fixing carbon dioxide in photosynthesis, some bacteria also fix nitrogen gas (nitrogen fixation) using the enzyme nitrogenase.

Growth and Reproduction

Unlike multicellular organisms, increases in the size of bacteria (cell growth) and their reproduction by cell division are tightly linked in unicellular organisms. Bacteria grow to a fixed size and then reproduce through binary fission, a form of asexual reproduction. Under optimal conditions, bacteria can grow and divide extremely rapidly, and bacterial populations can double as quickly as every 9.8 minutes. In cell division, two identical clone daughter cells are produced. Some bacteria, while still reproducing asexually, form more complex reproductive structures that help disperse the newly formed daughter cells. Examples include fruiting body formation by *Myxobacteria* and aerial hyphae formation by *Streptomyces*, or budding. Budding involves a cell forming a protrusion that breaks away and produces a daughter cell.

In the laboratory, bacteria are usually grown using solid or liquid media. Solid growth media such as agar plates are used to isolate pure cultures of a bacterial strain. However, liquid growth media are used when measurement of growth or large volumes of cells are required. Growth in stirred liquid media occurs as an even cell suspension, making the cultures easy to divide and transfer, although isolating single bacteria from liquid media is difficult. The use of selective media (media with specific nutrients added or deficient, or with antibiotics added) can help identify specific organisms.

Most laboratory techniques for growing bacteria use high levels of nutrients to produce large amounts of cells cheaply and quickly. However, in natural environments nutrients are limited, meaning that bacteria cannot continue to reproduce indefinitely. This nutrient limitation has led the evolution of different growth strategies (see r/ K selection theory).

Some organisms can grow extremely rapidly when nutrients become available, such as the formation of algal (and cyanobacterial) blooms that often occur in lakes during the summer. Other organisms have adaptations to harsh environments, such as the production of multiple antibiotics by *Streptomyces* that inhibit the growth of competing microorganisms. In nature, many organisms live in communities (e.g. biofilms) which may allow for increased supply of nutrients and protection from environmental stresses. These relationships can be essential for growth of a particular organism or group of organisms (syntrophy).

Bacterial growth follows three phases. When a population of bacteria first enter a high-nutrient environment that allows growth, the cells need to adapt to their new environment. The first phase of growth is the lag phase, a period of slow growth when the cells are adapting to the high-nutrient environment and preparing for fast growth.

The lag phase has high biosynthesis rates, as proteins necessary for rapid growth are produced. The second phase of growth is the logarithmic phase (log phase), also known as the exponential phase. The log phase is marked by rapid exponential growth. The rate at which cells grow during this phase is known as the *growth rate (k)*, and the time it takes the cells to double is known as the *generation time (g)*. During log phase, nutrients are metabolised at maximum speed until one of the nutrients is depleted and starts limiting growth. The final phase of growth is the *stationary phase* and is caused by depleted nutrients. The cells reduce their metabolic activity and consume non-essential cellular proteins. The stationary phase is a transition from rapid growth to a stress response state and there is increased expression of genes involved in DNA repair, antioxidant metabolism and nutrient transport.

Genetics

Most bacteria have a single circular chromosome that can range in size from only 160,000 base pairs in the endosymbiotic bacteria *Candidatus Carsonella ruddii*, to 12,200,000 base pairs in the soil-dwelling bacteria *Sorangium cellulosum*. Spirochaetes of the genus *Borrelia* are a notable exception to this arrangement, with bacteria such as *Borrelia burgdorferi*, the cause of Lyme disease, containing a single linear chromosome. The genes in bacterial genomes are usually a single continuous stretch of DNA and although several different types of introns do exist in bacteria, these are much more rare than in eukaryotes.

Bacteria may also contain plasmids, which are small extra-chromosomal DNAs that may contain genes for antibiotic resistance or virulence factors.

Bacteria, as asexual organisms, inherit identical copies of their parent's genes (i.e., they are clonal). However, all bacteria can evolve by selection on changes to their genetic material DNA caused by genetic recombination or mutations. Mutations come from errors made during the replication of DNA or from exposure to mutagens. Mutation rates vary widely among different species of bacteria and even among different clones of a single species of bacteria. Genetic changes in bacterial genomes come from either random mutation during replication or "stress-directed mutation", where genes involved in a particular growth-limiting process have an increased mutation rate. Some bacteria also transfer genetic material between cells. This can occur in three main ways. Firstly, bacteria can take up exogenous DNA from their environment, in a process called transformation. Genes can also be transferred by the process of transduction, when the integration of a bacteriophage introduces foreign DNA into the chromosome. The third method of gene transfer is bacterial conjugation, where DNA is transferred through direct cell contact. This gene acquisition from other bacteria or the environment is called horizontal gene transfer and may be common under natural conditions. Gene transfer is particularly important in antibiotic resistance as it allows the rapid transfer of resistance genes between different pathogens.

Bacteriophages

Bacteriophages are viruses that change the bacterial DNA. Many types of bacteriophage exist, some simply infect and lyse their host bacteria, while others insert into the bacterial chromosome. A bacteriophage can contain genes that contribute to its host's phenotype: for example, in the evolution of *Escherichia coli* O157:H7 and *Clostridium botulinum*, the toxin genes in an integrated phage converted a harmless ancestral bacteria into a lethal pathogen. Bacteria resist phage infection through restriction modification systems that degrade foreign DNA, and a system that uses CRISPR sequences to retain fragments of the genomes of phage that the bacteria have come into contact with in the past, which allows them to block virus replication through a form of RNA interference. This CRISPR system provides bacteria with acquired immunity to infection.

Movement

Motile bacteria can move using flagella, bacterial gliding, twitching motility or changes of buoyancy. In twitching motility, bacterial use their type IV pili as a grappling hook, repeatedly extending it, anchoring it and then retracting it with remarkable force (>80 pN).

Bacterial species differ in the number and arrangement of flagella on their surface; some have a single flagellum (monotrichous), a flagellum at each end (amphitrichous), clusters of flagella at the poles of the cell (lophotrichous), while others have flagella distributed over the entire surface of the cell (peritrichous). The bacterial flagella is the best-understood motility structure in any organism and is made of about 20 proteins, with approximately another 30 proteins required for its regulation and assembly. The flagellum is a rotating structure driven by a reversible motor at the base that uses the electrochemical gradient across the membrane for power. This motor drives the motion of the filament, which acts as a propeller.

Many bacteria (such as *E. coli*) have two distinct modes of movement: forward movement (swimming) and tumbling. The tumbling allows them to reorient and makes their movement a three-dimensional random walk. The flagella of a unique group of bacteria, the spirochaetes, are found between two membranes in the periplasmic space. They have a distinctive helical body that twists about as it moves. Motile bacteria are attracted or repelled by certain stimuli in behaviours called *taxes*: these include chemotaxis, phototaxis and magnetotaxis. In one peculiar group, the myxobacteria, individual bacteria move together to form waves of cells that then differentiate to form fruiting bodies containing spores. The myxobacteria move only when on solid surfaces, unlike *E. coli* which is motile in liquid or solid media.

Several *Listeria* and *Shigella* species move inside host cells by usurping the cytoskeleton, which is normally used to move organelles inside the cell. By promoting actin polymerization at one pole of their cells, they can form a kind of tail that pushes them through the host cell's cytoplasm.

Classification and Identification

Classification seeks to describe the diversity of bacterial species by naming and grouping organisms based on similarities. Bacteria can be classified on the basis of cell structure, cellular metabolism or on differences in cell components such as DNA, fatty acids, pigments, antigens and quinones. While these schemes allowed the identification and classification of bacterial strains, it was unclear whether these differences represented variation between distinct species or between strains of the same species. This uncertainty was due to the lack of distinctive structures in most bacteria, as well as lateral gene transfer between unrelated species.

Due to lateral gene transfer, some closely related bacteria can have very different morphologies and metabolisms. To overcome this uncertainty, modern bacterial classification emphasizes molecular systematics, using genetic techniques such as guanine cytosine ratio determination, genome-genome hybridization, as well as sequencing genes that have not undergone extensive lateral gene transfer, such as the rRNA gene. Classification of bacteria is determined by publication in the *International Journal of Systematic Bacteriology*, and Bergey's *Manual of Systematic Bacteriology*. The International Committee on Systematic Bacteriology (ICSB) maintains international rules for the naming of bacteria and taxonomic categories and for the ranking of them in the International Code of Nomenclature of Bacteria.

The term "bacteria" was traditionally applied to all microscopic, single-celled prokaryotes. However, molecular systematics showed prokaryotic life to consist of two separate domains, originally called *Eubacteria* and *Archaebacteria*, but now called *Bacteria* and *Archaea* that evolved independently from an ancient common ancestor. The archaea and eukaryotes are more closely related to each other than either is to the bacteria. These two domains, along with Eukarya, are the basis of the three-domain system, which is currently the most widely used classification system in microbiology. However, due to the relatively recent introduction of molecular systematics and a rapid increase in the number of genome sequences that are available, bacterial classification remains a changing and expanding field. For example, a few biologists argue that the Archaea and Eukaryotes evolved from Gram-positive bacteria.

Identification of bacteria in the laboratory is particularly relevant in medicine, where the correct treatment is determined by the bacterial species causing an infection. Consequently, the need to identify human pathogens was a major impetus for the development of techniques to identify bacteria.

The Gram stain, developed in 1884 by Hans Christian Gram, characterises bacteria based on the structural characteristics of their cell walls. The thick layers of peptidoglycan in the "Gram-positive" cell wall stain purple, while the thin "Gram-negative" cell wall appears pink. By combining morphology and Gram-staining, most bacteria can be classified as belonging to one of four groups (Gram-positive cocci, Gram-positive bacilli, Gram-negative cocci and Gram-negative bacilli). Some organisms are best identified by stains other than the Gram stain, particularly mycobacteria or *Nocardia*, which show acid-fastness on Ziehl–Neelsen or similar stains. Other organisms may need to be identified by their growth in special media, or by other techniques, such as serology.

Culture techniques are designed to promote the growth and identify particular bacteria, while restricting the growth of the other bacteria in the sample. Often these techniques are designed for specific specimens; for example, a sputum sample will be treated to identify organisms that cause pneumonia, while stool specimens are cultured on selective media to identify organisms that cause diarrhoea, while preventing growth of non-pathogenic bacteria. Specimens that are normally sterile, such as blood, urine or spinal fluid, are cultured under conditions designed to grow all possible organisms. Once a pathogenic organism has been isolated, it can be further characterised by its morphology, growth patterns such as (aerobic or anaerobic growth, patterns of hemolysis) and staining.

As with bacterial classification, identification of bacteria is increasingly using molecular methods. Diagnostics using such DNA-based tools, such as polymerase chain reaction, are increasingly popular due to their specificity and speed, compared to culture-based methods. These methods also allow the detection and identification of "viable but nonculturable" cells that are metabolically active but non-dividing. However, even using these improved methods, the total number of bacterial species is not known and cannot even be estimated with any certainty. Following present classification, there are fewer than 9,000 known species of bacteria (including cyanobacteria) , but attempts to estimate the true level of bacterial diversity have ranged from 10^7 to 10^9 total species - and even these diverse estimates may be off by many orders of magnitude.

Interactions with other Organisms

Despite their apparent simplicity, bacteria can form complex associations with other organisms. These symbiotic associations can be divided into parasitism, mutualism and commensalism. Due to their small size, commensal bacteria are ubiquitous and grow on animals and plants exactly as they will grow on any other surface. However, their growth can be increased by warmth and sweat, and large populations of these organisms in humans are the cause of body odor.

Predators

Some species of bacteria kill and then consume other microorganisms, these species called *predatory bacteria*. These include organisms such as *Myxococcus xanthus*, which forms swarms of cells that kill and digest any bacteria they encounter. Other bacterial predators either attach to their prey in order to digest them and absorb nutrients, such as *Vampirococcus*, or invade another cell and multiply inside the cytosol, such as *Daptobacter*. These predatory bacteria are thought to have evolved from saprophages that consumed dead microorganisms, through adaptations that allowed them to entrap and kill other organisms.

Mutualists

Certain bacteria form close spatial associations that are essential for their survival. One such mutualistic association, called interspecies hydrogen transfer, occurs between clusters of anaerobic bacteria that consume organic acids such as butyric acid or propionic acid and produce hydrogen, and methanogenic Archaea that consume hydrogen. The bacteria in this association are unable to consume the organic acids as this reaction produces hydrogen that accumulates in their surroundings. Only the intimate association with the hydrogen-consuming Archaea keeps the hydrogen concentration low enough to allow the bacteria to grow. In soil, microorganisms which reside in the rhizosphere (a zone that includes the root surface and the soil that adheres to the root after gentle shaking) carry out nitrogen fixation, converting nitrogen gas to nitrogenous compounds. This serves to provide an easily absorbable form of nitrogen for many plants, which cannot fix nitrogen themselves. Many other bacteria are found as symbionts in humans and other organisms. For example, the presence of over 1,000 bacterial species in the normal human gut flora of the intestines can contribute to gut immunity, synthesise vitamins such as folic acid, vitamin K and biotin, convert milk protein to lactic acid, as well as fermenting complex undigestible carbohydrates. The presence of this gut flora also inhibits the growth of potentially pathogenic bacteria (usually through competitive exclusion) and these beneficial bacteria are consequently sold as probiotic dietary supplements.

Pathogens

If bacteria form a parasitic association with other organisms, they are classed as pathogens. Pathogenic bacteria are a major cause of human death and disease and cause infections such as tetanus, typhoid fever, diphtheria, syphilis, cholera, foodborne illness, leprosy and tuberculosis.

A pathogenic cause for a known medical disease may only be discovered many years after, as was the case with Helicobacter pylori and peptic ulcer disease. Bacterial diseases are also important in agriculture, with bacteria causing leaf spot, fire blight and wilts in plants, as well as Johne's disease, mastitis, salmonella and anthrax in farm animals. Each species of pathogen has a characteristic spectrum of interactions with its human hosts. Some organisms, such as

Staphylococcus or *Streptococcus*, can cause skin infections, pneumonia, meningitis and even overwhelming sepsis, a systemic inflammatory response producing shock, massive vasodilation and death. Yet these organisms are also part of the normal human flora and usually exist on the skin or in the nose without causing any disease at all.

Other organisms invariably cause disease in humans, such as the Rickettsia, which are obligate intracellular parasites able to grow and reproduce only within the cells of other organisms.

One species of Rickettsia causes typhus, while another causes Rocky Mountain spotted fever. *Chlamydia*, another phylum of obligate intracellular parasites, contains species that can cause pneumonia, or urinary tract infection and may be involved in coronary heart disease. Finally, some species such as *Pseudomonas aeruginosa*, *Burkholderia cenocepacia*, and *Mycobacterium avium* are opportunistic pathogens and cause disease mainly in people suffering from immunosuppression or cystic fibrosis.

Bacterial infections may be treated with antibiotics, which are classified as bacteriocidal if they kill bacteria, or bacteriostatic if they just prevent bacterial growth. There are many types of antibiotics and each class inhibits a process that is different in the pathogen from that found in the host.

An example of how antibiotics produce selective toxicity are chloramphenicol and puromycin, which inhibit the bacterial ribosome, but not the structurally different eukaryotic ribosome. Antibiotics are used both in treating human disease and in intensive farming to promote animal growth, where they may be contributing to the rapid development of antibiotic resistance in bacterial populations.

Infections can be prevented by antiseptic measures such as sterilizating the skin prior to piercing it with the needle of a syringe, and by proper care of indwelling catheters. Surgical and dental instruments are also sterilized to prevent contamination by bacteria. Disinfectants such as bleach are used to kill bacteria or other pathogens on surfaces to prevent contamination and further reduce the risk of infection.

Significance in Technology and Industry

Bacteria, often lactic acid bacteria such as *Lactobacillus* and *Lactococcus*, in combination with yeasts and molds, have been used for thousands of years in the preparation of fermented foods such as cheese, pickles, soy sauce, sauerkraut, vinegar, wine and yoghurt.

The ability of bacteria to degrade a variety of organic compounds is remarkable and has been used in waste processing and bioremediation. Bacteria capable of digesting the hydrocarbons in petroleum are often used to clean up oil spills. Fertilizer was added to some of the beaches in Prince William Sound in an attempt to promote the growth of these naturally occurring bacteria after the infamous 1989 *Exxon Valdez* oil spill. These efforts were effective on beaches that were not too thickly covered in oil. Bacteria are also used for the bioremediation of industrial toxic wastes. In the chemical industry, bacteria are most important in the production of enantiomerically pure chemicals for use as pharmaceuticals or agrichemicals. Bacteria can also be used in the place of pesticides in the biological pest control. This commonly involves *Bacillus thuringiensis* (also called BT), a Gram-positive, soil dwelling bacterium. Subspecies of this bacteria are used as a Lepidopteran-specific insecticides under trade names such as Dipel and Thuricide. Because of their specificity, these pesticides are regarded as environmentally friendly, with little or no effect on humans, wildlife, pollinators and most other beneficial insects.

Because of their ability to quickly grow and the relative ease with which they can be manipulated, bacteria are the workhorses for the fields of molecular biology, genetics and biochemistry. By making mutations in bacterial DNA and examining the resulting phenotypes, scientists can determine the function of genes, enzymes and metabolic pathways in bacteria, then apply this knowledge to more complex organisms. This aim of understanding the biochemistry of a cell reaches its most complex expression in the synthesis of huge amounts of enzyme kinetic and gene expression data into mathematical models of entire organisms. This is achievable in some well-studied bacteria, with models of *Escherichia coli* metabolism now being produced and tested. This understanding of bacterial metabolism and genetics allows the use of biotechnology to bioengineer bacteria for the production of therapeutic proteins, such as insulin, growth factors, or antibodies.

Different Types of Bacteria

There are many types of bacteria which fall into different categories right from harmful to helpful bacteria and bacteria in different type of environments. Bacteria classification is mainly done based on morphology, condition required, DNA sequencing etc. Let us learn about the classification of bacteria.

What is bacteria? Bacteria is a single-celled organism which can only be seen through microscope. Bacteria comes in different shapes and the size of bacteria is measured in micrometer (which is a millionth part of a meter). Bacteria are found everywhere and in all type of environments.

There are numerous types of bacterial in the world. Before the invention of DNA sequencing technique, bacteria were mainly classified based on their shapes, which is also known as Morphology, biochemistry and staining (which is either Gram Positive or Gram Negative). Now a day along with the morphology, DNA sequencing is also used in order to classify bacteria. (DNA sequencing also helps in understanding relationship between two types of bacteria if they are related to each other despite of their shapes).

Along with the shape and DNA sequence, other things such as their metabolic activities, conditions required for their growth, biochemical reactions, antigenic properties, and other characteristics are also helpful in classifying the bacteria. There are various groups of bacteria, which belong to same family and are evolved from same bacteria (ancestor). However, each types of bacteria posses its own characteristics (those which are evolved after separation from the original specie).

Classification of Bacteria:

Bacteria are mainly classified into phylums (phylum is a scientific classification of organisms). For simplification, bacteria can be grouped into the following groups:

Bacteria classification based on shapes: As already mentioned, before the advent of DNA sequencing, bacteria were classified based on their shapes and biochemical properties. Most of the bacteria belong to three main shapes: rod (rod shaped bacteria are called bacilli), sphere (sphere shaped bacteria are called cocci) and spiral (spiral shaped bacteria are called spirilla). Some bacteria belong to different shapes.

Aerobic and anaerobic bacteria: Bacteria are also classified based on the requirement of oxygen for their survival. Bacteria those need oxygen for their survival are called Aerobic bacteria and bacteria those do not require oxygen for survival. Anaerobic bacteria cannot bear oxygen and may die if kept in oxygenated environment (anaerobic bacteria are found in places like under the surface of earth, deep ocean, and bacteria which live in some medium).

Gram Positive and Gram Negative bacteria: Bacteria are grouped as 'Gram Positive' bacteria and 'Gram Negative' bacteria, which is based on the results of Gram Staining Method (in which, an agent is used to bind to the cell wall of the bacteria) on bacteria.

Autotrophic and heterotrophic bacteria: This is one of the most important classification types as it takes into account the most important aspect of bacteria growth and reproduction. Autotrophic bacteria (also known as autotrophs) obtain the carbon it requires from carbon-dioxide. Some autotrophs directly use sun-light in order to produce sugar from carbon-dioxide whereas other depend on various chemical reactions. Heterotrophic bacteria obtain carob and/or sugar from the environment they are in (for example, the living cells or organism they are in).

Classification based on Phyla: Based on the morphology, DNA sequencing, conditions required and biochemistry, scientists have classified bacteria into phyla:

1) Aquificae
2) Xenobacteria
3) Fibrobacter
4) Bacteroids
5) Firmicutes
6) Planctomycetes
7) Chrysogenetic
8) Cyanobacteria
9) Thermomicrobia
10) Chlorobia
11) Proteobacteria
12) Spirochaetes
13) Flavobacteria
14) Fusobacteria
15) Verrucomicrobia.

Each phylum further corresponds to number of species and genera of bacteria. The bacteria classification includes bacteria which are found in various types of environments such as sweet water bacteria, ocean water bacteria, bacteria that can survive extreme temperatures (extreme hot as in sulfur water spring bacteria and extreme cold as in bacteria found in Antarctica ice), bacteria that can survive in highly acidic environment, bacteria that can survive highly alkaline environment, aerobic bacteria, anaerobic bacteria, autotrophic bacteria, heterotrophic bacteria, bacteria that can withstand high radiation etc.

Fungus

A fungus is any member of a large group of eukaryotic organisms that includes microorganisms such as yeasts and molds, as well as the more familiar mushrooms.

The Fungi are classified as a kingdom that is separate from plants and animals. One major difference is that fungal cells have cell walls that contain chitin, unlike the cell walls of plants, which contain cellulose.

These and other differences show that the fungi form a single group of related organisms, named the *Eumycota* (*true fungi* or *Eumycetes*), that share a common ancestor (a *monophyletic group*). This fungal group is distinct from the structurally similar slime molds (myxomycetes) and water molds (oomycetes). The discipline of biology devoted to the study of fungi is known as mycology, which is often regarded as a branch of botany, even though genetic studies have shown that fungi are more closely related to animals than to plants.

Abundant worldwide, most fungi are inconspicuous because of the small size of their structures, and their cryptic lifestyles in soil, on dead matter, and as symbionts of plants, animals, or other fungi. They may become noticeable when fruiting, either as mushrooms or molds. Fungi perform an essential role in the decomposition of organic matter and have fundamental roles in nutrient cycling and exchange.

They have long been used as a direct source of food, such as mushrooms and truffles, as a leavening agent for bread, and in fermentation of various food products, such as wine, beer, and soy sauce. Since the 1940s, fungi have been used for the production of antibiotics, and, more recently, various enzymes produced by fungi are used industrially and in detergents. Fungi are also used as biological agents to control weeds and pests.

Many species produce bioactive compounds called mycotoxins, such as alkaloids and polyketides, that are toxic to animals including humans. The fruiting structures of a few species are consumed recreationally or in traditional ceremonies as a source of psychotropic compounds. Fungi can break down manufactured materials and buildings, and become significant pathogens of humans and other animals. Losses of crops due to fungal diseases (e.g. rice blast disease) or food spoilage can have a large impact on human food supplies and local economies.

The fungus kingdom encompasses an enormous diversity of taxa with varied ecologies, life cycle strategies, and morphologies ranging from single-celled aquatic chytrids to large mushrooms. However, little is known of the true biodiversity of Kingdom Fungi, which has been estimated at around 1.5 million species, with about 5% of these having been formally classified. Ever since the pioneering 18th and 19th century taxonomical works of Carl Linnaeus, Christian Hendrik Persoon, and Elias Magnus Fries, fungi have been classified according to their morphology (e.g., characteristics such as spore colour or microscopic features) or physiology. Advances in molecular genetics have opened the way for DNA analysis to be incorporated into taxonomy, which has sometimes challenged the historical groupings based on morphology and other traits. Phylogenetic studies published in the last decade have helped reshape the classification of Kingdom Fungi, which is divided into one subkingdom, seven phyla, and ten subphyla.

Etymology

The English word *fungus* is directly adopted from the Latin *fungus* (mushroom), used in the writings of Horace and Pliny. This in turn is derived from the Greek word "sponge", which refers to the macroscopic structures and morphology of mushrooms and molds; the root is also used in other languages, such as the German *Schwamm* ("sponge"), *Schimmel* ("mold"), and the French *champignon* and the Spanish *champiñon* (which both mean "mushroom").

Characteristics

Before the introduction of molecular methods for phylogenetic analysis, taxonomists considered fungi to be members of the Plant Kingdom, based largely on similarities in lifestyle: both fungi and plants are mainly immobile, and have similarities in general morphology and growth habitat. Like plants, fungi often grow in soil, and in the case of mushrooms form conspicuous fruiting bodies, which sometimes bear resemblance to plants such as mosses. The fungi are now considered a separate kingdom, distinct from both plants and animals, from which they appear to have diverged around one billion years ago. Some morphological, biochemical, and genetic features are shared with other organisms, while others are unique to the fungi, clearly separating them from the other kingdoms:

Shared Features

- With other eukaryotes: As other eukaryotes, fungal cells contain membrane-bound nuclei with chromosomes that contain DNA with noncoding regions called introns and coding regions called exons. In addition, fungi possess membrane-bound cytoplasmic organelles such as mitochondria, sterol-containing membranes, and ribosomes of the 80S type. They have a characteristic range of soluble carbohydrates and storage compounds, including sugar alcohols (e.g., mannitol), disaccharides, (e.g., trehalose), and polysaccharides (e.g., glycogen, which is also found in animals).

- With animals: Fungi lack chloroplasts and are heterotrophic organisms, requiring preformed organic compounds as energy sources.

- With plants: Fungi possess a cell wall and vacuoles. They reproduce by both sexual and asexual means, and like basal plant groups (such as ferns and mosses) produce spores. Similar to mosses and algae, fungi typically have haploid nuclei.

- With euglenoids and bacteria: Higher fungi, euglenoids, and some bacteria produce the amino acid L-lysine in specific biosynthesis steps, called the á-aminoadipate pathway.

- The cells of most fungi grow as tubular, elongated, and thread-like structures and are called hyphae, which may contain multiple nuclei and extend at their tips. Each tip contains a set of aggregated vesicles—cellular structures consisting of proteins, lipids, and other organic molecules—called Spitzenkörper. Both fungi and oomycetes grow as filamentous hyphal cells. In contrast, similar-looking organisms, such as filamentous green algae, grow by repeated cell division within a chain of cells.

- In common with some plant and animal species, more than 60 fungal species display the phenomenon of bioluminescence.

Unique Features

- Some species grow as single-celled yeasts that reproduce by budding or binary fission. Dimorphic fungi can switch between a yeast phase and a hyphal phase in response to environmental conditions.

- The fungal cell wall is composed of glucans and chitin; while the former compounds are also found in plants and the latter in the exoskeleton of arthropods, fungi are the only organisms that combine these two structural molecules in their cell wall. In contrast to plants and the oomycetes, fungal cell walls do not contain cellulose.

Most fungi lack an efficient system for long-distance transport of water and nutrients, such as the xylem and phloem in many plants. To overcome these limitations, some fungi, such as *Armillaria*, form rhizomorphs, that resemble and perform functions similar to the roots of plants. Another characteristic shared with plants includes a biosynthetic pathway for producing terpenes that uses mevalonic acid and pyrophosphate as chemical building blocks. However, plants have an additional terpene pathway in their chloroplasts, a structure fungi do not possess. Fungi produce several secondary metabolites that are similar or identical in structure to those made by plants. Many of the plant and fungal enzymes that make these compounds differ from each other in sequence and other characteristics, which indicates separate origins and evolution of these enzymes in the fungi and plants.

Diversity

Fungi have a worldwide distribution, and grow in a wide range of habitats, including extreme environments such as deserts or areas with high salt concentrations or ionizing radiation, as well as in deep sea sediments. Some can survive the intense UV and cosmic radiation encountered during space travel. Most grow in terrestrial environments, but several species live partly or solely in aquatic habitats, such as the chytrid fungus *Batrachochytrium dendrobatidis*, a parasite that has been responsible for a worldwide decline in amphibian populations. This organism spends part of its life cycle as a motile zoospore, enabling it to propel itself through water and enter its amphibian host. Other examples of aquatic fungi include those living in hydrothermal areas of the ocean.

Around 100,000 species of fungi have been formally described by taxonomists, but the global biodiversity of the fungus kingdom is not fully understood. Based on observations of the ratio of the number of fungal species to the number of plant species in selected environments, the fungal kingdom has been estimated to contain about 1.5 million species. In mycology, species have historically been distinguished by a variety of methods and concepts. Classification based on morphological characteristics, such as the size and shape of spores or fruiting structures, has traditionally dominated fungal taxonomy. Species may also be distinguished by their biochemical and physiological characteristics, such as their ability to metabolize certain biochemicals, or their reaction to chemical tests. The biological species concept discriminates species based on their ability to mate. The application of molecular tools, such as DNA sequencing and phylogenetic analysis, to study diversity has greatly enhanced the resolution and added robustness to estimates of genetic diversity within various taxonomic groups.

Morphology

Microscopic Structures

Most fungi grow as hyphae, which are cylindrical, thread-like structures 2–10 µm in diameter and up to several centimetres in length. Hyphae grow at their tips (apices); new hyphae are typically formed by emergence of new tips along existing hyphae by a process called *branching*, or occasionally growing hyphal tips bifurcate (fork) giving rise to two parallel-growing hyphae.

The combination of apical growth and branching/forking leads to the development of a mycelium, an interconnected network of hyphae. Hyphae can be either septate or coenocytic: septate hyphae are divided into compartments separated by cross walls (internal cell walls, called septa, that are formed at right angles to the cell wall giving the hypha its shape), with each compartment containing one or more nuclei; coenocytic hyphae are not compartmentalized. Septa have pores, such as the dolipore septa in the basidiomycetes (which are fungi of the phylum Basidiomycota) that allow cytoplasm, organelles, and sometimes nuclei to pass through. Coenocytic hyphae are essentially multinucleate supercells.

Many species have developed specialized hyphal structures for nutrient uptake from living hosts; examples include haustoria in plant parasites of most fungal phyla, and arbuscules of several mycorrhizal fungi, which penetrate into the host cells to consume nutrients.

Although fungi are part of the opisthokont clade—a grouping of evolutionarily-related organisms broadly characterized by a single posterior flagellum—all phyla except for the chytrids have lost their posterior flagella. Fungi are unusual among the eukaryotes in having a cell wall that, in addition to glucans (e.g., â-1,3-glucan) and other typical components, also contains the biopolymer chitin.

Macroscopic Structures

Fungal mycelia can become visible macroscopically, for example, as concentric rings on various surfaces, such as damp walls, and on other substrates, such as spoilt food, and are commonly and generically called mold (British spelling, mould); mycelia grown on solid agar media in laboratory petri dishes are usually referred to as colonies, exhibiting characteristic macroscopic growth shapes and colours, due to spores or pigmentation. Some individual fungal colonies can grow to a very large size and mass, in some cases reaching extraordinary dimensions and ages as in the case of a clonal colony of *Armillaria ostoyae*, which extends over an area of more than 900 ha, with an estimated age of nearly 9,000 years.

In the ascomycetes, a specialized structure important in sexual reproduction is the apothecium, a cup-shaped structure that holds the hymenium, a layer of tissue containing the spore-bearing cells. The fruiting bodies of the basidiomycetes and some ascomycetes can sometimes grow very large, and are well-known as mushrooms.

Growth and Physiology

The growth of fungi as filamentous hyphae on or in solid substrates or as single cells in aquatic environments is adapted for the efficient extraction of nutrients, because these growth forms have high surface area to volume ratios. Hyphae are specifically adapted for growth on solid surfaces, and to invade substrates and tissues. They can exert large penetrative mechanical forces; for example, the plant pathogen *Magnaporthe grisea* forms a structure called an appressorium which evolved to puncture plant tissues.

The pressure generated by the appressorium, directed against the plant epidermis, can exceed 8 MPa (80 bars). The filamentous fungus *Paecilomyces lilacinus* uses a similar structure to penetrate the eggs of plant-parasitic nematodes.

Fungal reproduction is complex, reflecting the heterogeneity in lifestyles and genetic makeup within this Kingdom of organisms. It is estimated that a third of all fungi use more than one type of reproduction, frequently in two well-differentiated life cycle stages (the teleomorph and the anamorph). Environmental conditions trigger genetically determined developmental programs that lead to the creation of specialized structures for sexual or asexual reproduction. These structures aid both reproduction and efficient dispersal of spores or spore-containing propagules.

Asexual Reproduction

Asexual reproduction via vegetative spores (conidia) or through mycelial fragmentation is common; it maintains clonal populations adapted to a specific niche, and allows more rapid dispersal than sexual reproduction. The "Fungi imperfecti" (fungi lacking the perfect or sexual stage) or Deuteromycota comprise all the species which lack an observable sexual cycle.

Sexual Reproduction

Sexual reproduction with meiosis exists in all fungal phyla (with the exception of the Glomeromycota). It differs in many aspects from sexual reproduction in animals or plants. Differences also exist between fungal groups and can be used to discriminate species based on morphological differences in sexual structures and reproductive strategies. Mating experiments between fungal isolates may identify species based on biological species concepts. The major fungal groupings have initially been delineated based on the morphology of their sexual structures and spores; for example, the spore-containing structures, asci and basidia, can be used in the identification of ascomycetes and basidiomycetes, respectively. Some species may allow mating only between individuals of opposite mating type, while others can mate and sexually reproduce with any other individual or itself. Species of the former mating system are called heterothallic, and of the latter homothallic.

Most fungi have both a haploid and diploid stage in their life cycles. In sexually reproducing fungi, compatible individuals may combine by fusing their hyphae together into an interconnected network; this process, anastomosis, is required for the initiation of the sexual cycle. Ascomycetes and basidiomycetes go through a dikaryotic stage, in which the nuclei inherited from the two parents do not combine immediately after cell fusion, but remain separate in the hyphal cells.

In ascomycetes, dikaryotic hyphae of the hymenium (the spore-bearing tissue layer) form a characteristic *hook* at the hyphal septum. During cell division, formation of the hook ensures proper distribution of the newly divided nuclei into the apical and basal hyphal compartments. An ascus (plural *asci*) is then formed, in which karyogamy (nuclear fusion) occurs. Asci are embedded in an ascocarp, or fruiting body. Karyogamy in the asci is followed immediately by meiosis and the production of ascospores. After dispersal, the ascospores may germinate and form a new haploid mycelium. Sexual reproduction in basidiomycetes is similar to that of the ascomycetes. Compatible haploid hyphae fuse to produce a dikaryotic mycelium. However, the dikaryotic phase is more extensive in the basidiomycetes, often also present in the vegetatively growing mycelium. A specialized anatomical structure, called a clamp connection, is formed at each hyphal septum.

As with the structurally similar hook in the ascomycetes, the clamp connection in the basidiomycetes is required for controlled transfer of nuclei during cell division, to maintain the dikaryotic stage with two genetically different nuclei in each hyphal compartment. A basidiocarp is formed in which club-like structures known as basidia generate haploid basidiospores after karyogamy and meiosis. The most commonly known basidiocarps are mushrooms, but they may also take other forms. In glomeromycetes (formerly zygomycetes), haploid hyphae of two individuals fuse, forming a gametangium, a specialized cell structure that becomes a fertile gamete-producing cell. The gametangium develops into a zygospore, a thick-walled spore formed by the union of gametes. When the zygospore germinates, it undergoes meiosis, generating new haploid hyphae, which may then form asexual sporangiospores. These sporangiospores allow the fungus to rapidly disperse and germinate into new genetically identical haploid fungal mycelia.

Spore Dispersal

Both asexual and sexual spores or sporangiospores are often actively dispersed by forcible ejection from their reproductive structures. This ejection ensures exit of the spores from the reproductive structures as well as travelling through the air over long distances.

Specialized mechanical and physiological mechanisms as well as spore-surface structures, such as hydrophobins, enable efficient spore ejection. For example, the structure of the spore-bearing cells in some ascomycete species is such that the buildup of substances affecting cell volume and fluid balance enables the explosive discharge of spores into the air.

The forcible discharge of single spores termed *ballistospores* involves formation of a small drop of water (Buller's drop), which upon contact with the spore leads to its projectile release with an initial acceleration of more than 10,000 g; the net result is that the spore is ejected 0.01–0.02 cm, sufficient distance for it to fall through the gills or pores into the air below. Other fungi, like the puffballs, rely on alternative mechanisms for spore release, such as external mechanical forces. The bird's nest fungi use the force of falling water drops to liberate the spores from cup-shaped fruiting bodies. Another strategy is seen in the stinkhorns, a group of fungi with lively colours and putrid odor that attract insects to disperse their spores.

Other Sexual Processes

Besides regular sexual reproduction with meiosis, certain fungi, such as those in the genera *Penicillium* and *Aspergillus*, may exchange genetic material via parasexual processes, initiated by anastomosis between hyphae and plasmogamy of fungal cells. The frequency and relative importance of parasexual events is unclear and may be lower than other sexual processes. It is known to play a role in intraspecific hybridization and is likely required for hybridization between species, which has been associated with major events in fungal evolution.

Evolution

In contrast to plants and animals, the early fossil record of the fungi is meager. Factors that likely contribute to the under-representation of fungal species among fossils include the nature of fungal fruiting bodies, which are soft, fleshy, and easily degradable tissues and the microscopic dimensions of most fungal structures, which therefore are not readily evident. Fungal fossils are difficult to distinguish from those of other microbes, and are most easily identified when they resemble extant fungi. Often recovered from a permineralized plant or animal host, these samples are typically studied by making thin-section preparations that can be examined with light microscopy or transmission electron microscopy. Compression fossils are studied by dissolving the surrounding matrix with acid and then using light or scanning electron microscopy to examine surface details.

The earliest fossils possessing features typical of fungi date to the Proterozoic eon, some 1,430 million years ago (Ma); these multicellular benthic organisms had filamentous structures with septa, and were capable of anastomosis. More recent studies (2009) estimate the arrival of fungal organisms at about 760–1060 Ma based on comparisons of the rate of evolution in closely related groups. For much of the Paleozoic Era (542–251 Ma), the fungi appear to have been aquatic and consisted of organisms similar to the extant Chytrids in having flagellum-bearing spores. The evolutionary adaptation from an aquatic to a terrestrial lifestyle necessitated a diversification of ecological strategies for obtaining nutrients, including parasitism, saprobism, and the development of mutualistic relationships such as mycorrhiza and lichenization. Recent (2009) studies suggest that the ancestral ecological state of the Ascomycota was saprobism, and that independent lichenization events have occurred multiple times.

The fungi probably colonized the land during the Cambrian (542–488.3 Ma), long before land plants. Fossilized hyphae and spores recovered from the Ordovician of Wisconsin (460 Ma) resemble modern-day Glomerales, and existed at a time when the land flora likely consisted of only non-vascular bryophyte-like plants. Fungal fossils do not become common and uncontroversial until the early Devonian (416–359.2 Ma), when they are abundant in the Rhynie chert, mostly as Zygomycota and Chytridiomycota. At about this same time, approximately 400 Ma, the Ascomycota and Basidiomycota diverged, and all modern classes of fungi were present by the Late Carboniferous (Pennsylvanian, 318.1–299 Ma).

Lichen-like fossils have been found in the Doushantuo Formation in southern China dating back to 635–551 Ma. Lichens were a component of the early terrestrial ecosystems, and the estimated age of the oldest terrestrial lichen fossil is 400 Ma; this date corresponds to the age of the oldest known sporocarp fossil, a *Paleopyrenomycites* species found in the Rhynie Chert. The oldest fossil with microscopic features resembling modern-day basidiomycetes is *Palaeoancistrus*, found permineralized with a fern from the Pennsylvanian. Rare in the fossil record are the homobasidiomycetes (a taxon roughly equivalent to the mushroom-producing species of the agaricomycetes). Based on two amber-preserved specimens, the earliest known mushroom-forming fungi (the extinct species *Archaeomarasmius legletti*) appeared during the mid-Cretaceous, 90 Ma. Some time after the Permian-Triassic extinction event (251.4 Ma), a fungal spike (originally thought to be an extraordinary abundance of fungal spores in sediments) formed, suggesting that fungi were the dominant life form at this time, representing nearly 100% of the available fossil record for this period. However, the relative proportion of fungal spores relative to spores formed by algal species is difficult to assess, the spike did not appear worldwide, and in many places it did not fall on the Permian-Triassic boundary.

Taxonomy

Even though traditionally included in many botany curricula and textbooks, fungi are now thought to be more closely related to animals than to plants and are placed with the animals in the monophyletic group of opisthokonts. Analyses using molecular phylogenetics support a monophyletic origin of the Fungi. The taxonomy of the Fungi is in a state of constant flux, especially due to recent research based on DNA comparisons. These current phylogenetic analyses often overturn classifications based on older and sometimes less discriminative methods based on morphological features and biological species concepts obtained from experimental matings.

There is no unique generally accepted system at the higher taxonomic levels and there are frequent name changes at every level, from species upwards. Efforts among researchers are now underway to establish and encourage usage of a unified and more consistent nomenclature. Fungal species can also have multiple scientific names depending on their life cycle and mode (sexual or asexual) of reproduction. Web sites such as Index Fungorum and ITIS list current names of fungal species (with cross-references to older synonyms).

The 2007 classification of Kingdom Fungi is the result of a large-scale collaborative research effort involving dozens of mycologists and other scientists working on fungal taxonomy. It recognizes seven phyla, two of which—the Ascomycota and the Basidiomycota—are contained within a branch representing subkingdom Dikarya. The below cladogram depicts the major fungal taxa and their relationship to opisthokont and unikont organisms. The lengths of the branches in this tree are not proportional to evolutionary distances.

Taxonomic Groups

The major phyla (sometimes called divisions) of fungi have been classified based mainly on the characteristics of their sexual reproductive structures. Currently, seven phyla are proposed: Microsporidia, Chytridiomycota, Blastocladiomycota, Neocallimastigomycota, Neocallimastigomycota, Glomeromycota, Ascomycota, and Basidiomycota.

Phylogenetic analysis has demonstrated that the Microsporidia, unicellular parasites of animals and protists, are fairly recent and highly derived endobiotic fungi (living within the tissue of another species). One 2006 study concludes that the Microsporidia are a sister group to the true fungi, that is, they are each other's closest evolutionary relative. Hibbett and colleagues suggest that this analysis does not clash with their classification of the Fungi, and although the Microsporidia are elevated to phylum status, it is acknowledged that further analysis is required to clarify evolutionary relationships within this group. The Chytridiomycota are commonly known as chytrids. These fungi are distributed worldwide. Chytrids produce zoospores that are capable of active movement through aqueous phases with a single flagellum, leading early taxonomists to classify them as protists. Molecular phylogenies, inferred from rRNA sequences in ribosomes, suggest that the Chytrids are a basal group divergent from the other fungal phyla, consisting of four major clades with suggestive evidence for paraphyly or possibly polyphyly.

The Blastocladiomycota were previously considered a taxonomic clade within the Chytridiomycota. Recent molecular data and ultrastructural characteristics, however, place the Blastocladiomycota as a sister clade to the Zygomycota, Glomeromycota, and Dikarya (Ascomycota and Basiomycota). The blastocladiomycetes are saprotrophs, feeding on decomposing organic matter, and they are parasites of all eukaryotic groups. Unlike their close relatives, the chytrids, which mostly exhibit zygotic meiosis, the blastocladiomycetes undergo sporic meiosis.

The Neocallimastigomycota were earlier placed in the phylum Chytridomycota. Members of this small phylum are anaerobic organisms, living in the digestive system of larger herbivorous mammals and possibly in other terrestrial and aquatic environments. They lack mitochondria but contain hydrogenosomes of mitochondrial origin. As the related chrytrids, neocallimastigomycetes form zoospores that are posteriorly uniflagellate or polyflagellate.

Members of the Glomeromycota form arbuscular mycorrhizae, a form of symbiosis where fungal hyphae invade plant root cells and both species benefit from the resulting increased supply of nutrients. All known Glomeromycota species reproduce asexually.

The symbiotic association between the Glomeromycota and plants is ancient, with evidence dating to 400 million years ago. Formerly part of the Zygomycota (commonly known as 'sugar' and 'pin' molds), the Glomeromycota were elevated to phylum status in 2001 and now replace the older phylum Zygomycota. Fungi that were placed in the Zygomycota are now being reassigned to the Glomeromycota, or the subphyla incertae sedis Mucoromycotina, Kickxellomycotina, the Zoopagomycotina and the Entomophthoromycotina. Some well-known examples of fungi formerly in the Zygomycota include black bread mold (*Rhizopus stolonifer*), and *Pilobolus* species, capable of ejecting spores several meters through the air. Medically relevant genera include *Mucor, Rhizomucor*, and *Rhizopus*. The Ascomycota, commonly known as sac fungi or ascomycetes, constitute the largest taxonomic group within the Eumycota. These fungi form meiotic spores called ascospores, which are enclosed in a special sac-like structure called an ascus.

This phylum includes morels, a few mushrooms and truffles, single-celled yeasts (e.g., of the genera *Saccharomyces, Kluyveromyces, Pichia*, and *Candida*), and many filamentous fungi living as saprotrophs, parasites, and mutualistic symbionts. Prominent and important genera of filamentous ascomycetes include *Aspergillus, Penicillium, Fusarium*, and *Claviceps*. Many ascomycete species have only been observed undergoing asexual reproduction (called anamorphic species), but analysis of molecular data has often been able to identify their closest teleomorphs in the Ascomycota. Because the products of meiosis are retained within the sac-like ascus, ascomycetes have been used for elucidating principles of genetics and heredity (e.g. *Neurospora crassa*).

Members of the Basidiomycota, commonly known as the club fungi or basidiomycetes, produce meiospores called basidiospores on club-like stalks called basidia. Most common mushrooms belong to this group, as well as rust and smut fungi, which are major pathogens of grains. Other important basidiomycetes include the maize pathogen *Ustilago maydis*, human commensal species of the genus *Malassezia*, and the opportunistic human pathogen, *Cryptococcus neoformans*.

Fungus-like Organisms

Because of similarities in morphology and lifestyle, the slime molds (myxomycetes) and water molds (oomycetes) were formerly classified in the kingdom Fungi. Unlike true fungi the cell walls of these organisms contain cellulose and lack chitin. Slime molds are unikonts like fungi, but are grouped in the Amoebozoa. Water molds are diploid bikonts, grouped in the Chromalveolate kingdom. Neither water molds nor slime molds are closely related to the true fungi, and, therefore, taxonomists no longer group them in the kingdom Fungi. Nonetheless, studies of the oomycetes and myxomycetes are still often included in mycology textbooks and primary research literature.

The nucleariids, currently grouped in the Choanozoa, may be a sister group to the eumycete clade, and as such could be included in an expanded fungal kingdom.

Ecology

Although often inconspicuous, fungi occur in every environment on Earth and play very important roles in most ecosystems. Along with bacteria, fungi are the major decomposers in most terrestrial (and some aquatic) ecosystems, and therefore play a critical role in biogeochemical cycles and in many food webs. As decomposers, they play an essential role in nutrient cycling, especially as saprotrophs and symbionts, degrading organic matter to inorganic molecules, which can then re-enter anabolic metabolic pathways in plants or other organisms.

Symbiosis

Many fungi have important symbiotic relationships with organisms from most if not all Kingdoms. These interactions can be mutualistic or antagonistic in nature, or in the case of commensal fungi are of no apparent benefit or detriment to the host.

With Plants

Mycorrhizal symbiosis between plants and fungi is one of the most well-known plant–fungus associations and is of significant importance for plant growth and persistence in many ecosystems; over 90% of all plant species engage in mycorrhizal relationships with fungi and are dependent upon this relationship for survival.

The mycorrhizal symbiosis is ancient, dating to at least 400 million years ago. It often increases the plant's uptake of inorganic compounds, such as nitrate and phosphate from soils having low concentrations of these key plant nutrients. The fungal partners may also mediate plant-to-plant transfer of carbohydrates and other nutrients. Such mycorrhizal communities are called "common mycorrhizal networks". Some species inhabit the tissues inside roots, stems, and leaves, in which case they are called endophytes. Similar to mycorrhiza, endophytic colonization by fungi may benefit both symbionts; for example, endophytes of grasses impart to their host increased resistance to herbivores and other environmental stresses and receive food and shelter from the plant in return.

With Algae and Cyanobacteria

Lichens are formed by a symbiotic relationship between algae or cyanobacteria (referred to in lichen terminology as "photobionts") and fungi (mostly various species of ascomycetes and a few basidiomycetes), in which individual photobiont cells are embedded in a tissue formed by the fungus. Lichens occur in every ecosystem on all continents, play a key role in soil formation and the initiation of biological succession, and are the dominating life forms in extreme environments, including polar, alpine, and semiarid desert regions. They are able to grow on inhospitable surfaces, including bare soil, rocks, tree bark, wood, shells, barnacles and leaves. As in mycorrhizas, the photobiont provides sugars and other carbohydrates via photosynthesis, while the fungus provides minerals and water. The functions of both symbiotic organisms are so closely intertwined that they function almost as a single organism; in most cases the resulting organism differs greatly from the individual components. Lichenization is a common mode of nutrition; around 20% of fungi—between 17,500 and 20,000 described species—are lichenized. Characteristics common to most lichens include obtaining organic carbon by photosynthesis, slow growth, small size, long life, long-lasting (seasonal) vegetative reproductive structures, mineral nutrition obtained largely from airborne sources, and greater tolerance of dessication than most other photosynthetic organisms in the same habitat.

With Insects

Many insects also engage in mutualistic relationships with fungi. Several groups of ants cultivate fungi in the order Agaricales as their primary food source, while ambrosia beetles cultivate various species of fungi in the bark of trees that they infest. Similarly, females of several wood wasp species (genus *Sirex*) inject their eggs together with spores of the wood-rotting fungus *Amylostereum areolatum* into the sapwood of pine trees; the growth of the fungus provides ideal nutritional conditions for the development of the wasp larvae. Termites on the African savannah are also known to cultivate fungi, and yeasts of the genera *Candida* and *Lachancea* inhabit the gut of a wide range of insects, including neuropterans, beetles, and cockroaches; it is not known whether these fungi benefit their hosts.

As Pathogens and Parasites

Many fungi are parasites on plants, animals (including humans), and other fungi. Serious pathogens of many cultivated plants causing extensive damage and losses to agriculture and forestry include the rice blast fungus *Magnaporthe oryzae*, tree pathogens such as *Ophiostoma ulmi* and *Ophiostoma novo-ulmi* causing Dutch elm disease, and *Cryphonectria parasitica* responsible for chestnut blight, and plant pathogens in the genera *Fusarium, Ustilago, Alternaria,* and *Cochliobolus*. Some carnivorous fungi, like *Paecilomyces lilacinus*, are predators of nematodes, which they capture using an array of specialized structures such as constricting rings or adhesive nets.

Some fungi can cause serious diseases in humans, several of which may be fatal if untreated. These include aspergilloses, candidoses, coccidioidomycosis, cryptococcosis, histoplasmosis, mycetomas, and paracoccidioidomycosis. Furthermore, persons with immuno-deficiencies are particularly susceptible to disease by genera such as *Aspergillus*, *Candida*, *Cryptoccocus*, *Histoplasma*, and *Pneumocystis*. Other fungi can attack eyes, nails, hair, and especially skin, the so-called dermatophytic and keratinophilic fungi, and cause local infections such as ringworm and athlete's foot. Fungal spores are also a cause of allergies, and fungi from different taxonomic groups can evoke allergic reactions.

Human Use

The human use of fungi for food preparation or preservation and other purposes is extensive and has a long history. Mushroom farming and mushroom gathering are large industries in many countries. The study of the historical uses and sociological impact of fungi is known as ethnomycology. Because of the capacity of this group to produce an enormous range of natural products with antimicrobial or other biological activities, many species have long been used or are being developed for industrial production of antibiotics, vitamins, and anti-cancer and cholesterol-lowering drugs. More recently, methods have been developed for genetic engineering of fungi, enabling metabolic engineering of fungal species. For example, genetic modification of yeast species —which are easy to grow at fast rates in large fermentation vessels—has opened up ways of pharmaceutical production that are potentially more efficient than production by the original source organisms.

Antibiotics

Many species produce metabolites that are major sources of pharmacologically active drugs. Particularly important are the antibiotics, including the penicillins, a structurally related group of â-lactam antibiotics that are synthesized from small peptides. Although naturally occurring penicillins such as penicillin G (produced by *Penicillium chrysogenum*) have a relatively narrow spectrum of biological activity, a wide range of other penicillins can be produced by chemical modification of the natural penicillins. Modern penicillins are semisynthetic compounds, obtained initially from fermentation cultures, but then structurally altered for specific desirable properties. Other antibiotics produced by fungi include: griseofulvin from *Penicillium griseofulvin* used to treat dermatophyte infections of the skin, hair and nails; cyclosporins, commonly used as an immunosuppressant during transplant surgery; and fusidic acid, used to help control infection from methicillin-resistant *Staphylococcus aureus* bacteria. Widespread use of these antibiotics for the treatment of bacterial diseases, such as tuberculosis, syphilis, leprosy, and many others began in the early 20th century and continues to play a major part in anti-bacterial chemotherapy. In nature, antibiotics of fungal or bacterial origin appear to play a dual role: at high concentrations they act as chemical defence against competition with other microorganisms in species-rich environments, such as the rhizosphere, and at low concentrations as quorum-sensing molecules for intra- or interspecies signaling.

Cultured foods

Baker's yeast or *Saccharomyces cerevisiae*, a single-celled fungus, is used to make bread and other wheat-based products, such as pizza dough and dumplings. Yeast species of the genus *Saccharomyces* are also used to produce alcoholic beverages through fermentation. Shoyu koji mold (*Aspergillus oryzae*) is an essential ingredient in brewing Shoyu (soy sauce) and sake, and the preparation of miso, while *Rhizopus* species are used for making tempeh. Several of these fungi are domesticated species that were bred or selected based on their capacity to ferment food without producing harmful mycotoxins, which are produced by very closely related *Aspergilli*. Quorn, a meat substitute, is made from *Fusarium venenatum*.

Medicinal Use

Certain mushrooms enjoy usage as therapeutics in traditional and folk medicines, such as Traditional Chinese medicine. Notable medicinal mushrooms with a well-documented history of use include *Agaricus blazei, Ganoderma lucidum*, and *Cordyceps sinensis*. Research has identified compounds produced by these and other fungi that have inhibitory biological effects against viruses and cancer cells. Specific metabolites with biological or antimicrobial activities, such as polysaccharide-K, ergotamine, and â-lactam antibiotics, are routinely used in clinical medicine. The shiitake mushroom is a source of lentinan, a clinical drug approved for use in cancer treatments in several countries, including Japan. In Europe and Japan, polysaccharide-K (brand name Krestin), a chemical derived from *Trametes versicolor*, is an approved adjuvant for cancer therapy.

Edible and Poisonous Species

Edible mushrooms are well-known examples of fungi. Many are commercially raised, but others must be harvested from the wild. *Agaricus bisporus*, sold as button mushrooms when small or Portobello mushrooms when larger, is a commonly-eaten species, used in salads, soups, and many other dishes. Many Asian fungi are commercially grown and have increased in popularity in the West. They are often available fresh in grocery stores and markets, including straw mushrooms (*Volvariella volvacea*), oyster mushrooms (*Pleurotus ostreatus*), shiitakes (*Lentinula edodes*), and enokitake (*Flammulina* spp.).

There are many more mushroom species that are harvested from the wild for personal consumption or commercial sale. Milk mushrooms, morels, chanterelles, truffles, black trumpets, and *porcini* mushrooms (*Boletus edulis*) (also known as king boletes) demand a high price on the market. They are often used in gourmet dishes.

Certain types of cheeses require inoculation of milk curds with fungal species that impart a unique flavor and texture to the cheese. Examples include the blue colour in cheeses such as Stilton or Roquefort, which are made by inoculation with *Penicillium roqueforti*. Molds used in cheese production are non-toxic and are thus safe for human consumption; however, mycotoxins (e.g., aflatoxins, roquefortine C, patulin, or others) may accumulate due to growth of other fungi during cheese ripening or storage.

Many mushroom species are poisonous to humans, with toxicities ranging from slight digestive problems or allergic reactions as well as hallucinations to severe organ failures and death. Genera with mushrooms containing deadly toxins include *Conocybe*, *Galerina*, *Lepiota*, and most infamously, *Amanita*. The latter genus includes the destroying angel *(A. virosa)* and the death cap *(A. phalloides)*, the most common cause of deadly mushroom poisoning. The false morel (*Gyromitra esculenta*) is occasionally considered a delicacy when cooked, yet can be highly toxic when eaten raw. *Tricholoma equestre* was considered edible until being implicated in serious poisonings causing rhabdomyolysis. Fly agaric mushrooms (*Amanita muscaria*) also cause occasional non-fatal poisonings, mostly as a result of ingestion for use as a recreational drug for its hallucinogenic properties. Historically, fly agaric was used by different peoples in Europe and Asia and its present usage for religious or shamanic purposes is reported from some ethnic groups such as the Koryak people of north-eastern Siberia.

As it is difficult to accurately identify a safe mushroom without proper training and knowledge, it is often advised to assume that a wild mushroom is poisonous and not to consume it.

Pest Control

In agriculture, fungi may be useful if they actively compete for nutrients and space with pathogenic microorganisms such as bacteria or other fungi via the competitive exclusion principle, or if they are parasites of these pathogens.

For example, certain species may be used to eliminate or suppress the growth of harmful plant pathogens, such as insects, mites, weeds, nematodes and other fungi that cause diseases of important crop plants. This has generated strong interest in practical applications that use these fungi in the biological control of these agricultural pests.

Entomopathogenic fungi can be used as biopesticides, as they actively kill insects. Examples that have been used as biological insecticides are *Beauveria bassiana, Metarhizium anisopliae, Hirsutella* spp, *Paecilomyces* spp, and *Verticillium lecanii*. Endophytic fungi of grasses of the genus *Neotyphodium*, such as *N. coenophialum*, produce alkaloids that are toxic to a range of invertebrate and vertebrate herbivores. These alkaloids protect grass plants from herbivory, but several endophyte alkaloids can poison grazing animals, such as cattle and sheep. Infecting cultivars of pasture or forage grasses with *Neotyphodium* endophytes is one approach being used in grass breeding programs; the fungal strains are selected for producing only alkaloids that increase resistance to herbivores such as insects, while being non-toxic to livestock.

Bioremediation

Certain fungi, in particular "white rot" fungi, can degrade insecticides, herbicides, pentachlorophenol, creosote, coal tars, and heavy fuels and turn them into carbon dioxide, water, and basic elements. Fungi have been shown to biomineralize uranium oxides, suggesting they may have application in the bioremediation of radioactively polluted sites.

Model Organisms

Several pivotal discoveries in biology were made by researchers using fungi as model organisms, that is, fungi that grow and sexually reproduce rapidly in the laboratory. For example, the one gene-one enzyme hypothesis was formulated by scientists who used the bread mold *Neurospora crassa* to test their biochemical theories. Other important model fungi are *Aspergillus nidulans* and the yeasts, *Saccaromyces cerevisiae* and *Schizosaccharomyces pombe*, each of which has a long history of use to investigate issues in eukaryotic cell biology and genetics, such as cell cycle regulation, chromatin structure, and gene regulation. Other fungal models have more recently emerged that each address specific biological questions relevant to medicine, plant pathology, and industrial uses; examples include *Candida albicans*, a dimorphic, opportunistic human pathogen, *Magnaporthe grisea*, a plant pathogen, and *Pichia pastoris*, a yeast widely used for eukaryotic protein expression.

Others

Fungi are used extensively to produce industrial chemicals like citric, gluconic, lactic, and malic acids, antibiotics, and even to make stonewashed jeans. Fungi are also sources of industrial enzymes, such as lipases used in biological detergents, amylases, cellulases, invertases, proteases and xylanases. Several species, most notably *Psilocybin mushrooms* (colloquially known as *magic mushrooms*), are ingested for their psychedelic properties, both recreationally and religiously.

Mycotoxins

Many fungi produce biologically active compounds, several of which are toxic to animals or plants and are therefore called mycotoxins. Of particular relevance to humans are mycotoxins produced by molds causing food spoilage, and poisonous mushrooms. Particularly infamous are the lethal amatoxins in some *Amanita* mushrooms, and ergot alkaloids, which have a long history of causing serious epidemics of ergotism (St Anthony's Fire) in people consuming rye or related cereals contaminated with sclerotia of the ergot fungus, *Claviceps purpurea*. Other notable mycotoxins include the aflatoxins, which are insidious liver toxins and highly carcinogenic metabolites produced by certain *Aspergillus* species often growing in or on grains and nuts consumed by humans, ochratoxins, patulin, and trichothecenes (e.g., T-2 mycotoxin) and fumonisins, which have significant impact on human food supplies or animal livestock.

Mycotoxins are secondary metabolites (or natural products), and research has established the existence of biochemical pathways solely for the purpose of producing mycotoxins and other natural products in fungi. Mycotoxins may provide fitness benefits in terms of physiological adaptation, competition with other microbes and fungi, and protection from consumption (fungivory).

Mycology

Mycology is the branch of biology concerned with the systematic study of fungi, including their genetic and biochemical properties, their taxonomy, and their use to humans as a source of medicine, food, and psychotropic substances consumed for religious purposes, as well as their dangers, such as poisoning or infection. The field of phytopathology, the study of plant diseases, is closely related because many plant pathogens are fungi.

Use of fungi by humans dates back to prehistory; Ötzi the Iceman, a well-preserved mummy of a 5,300 year old Neolithic man found frozen in the Austrian Alps, carried two species of polypore mushrooms that may have been used as tinder (*Fomes fomentarius*), or for medicinal purposes (*Piptoporus betulinus*). Ancient peoples have used fungi as food sources – often unknowingly – for millennia, in the preparation of leavened bread and fermented juices. Some of the oldest written records contain references to the destruction of crops that were probably caused by pathogenic fungi.

History

Mycology is a relatively new science that became systematic after the development of the microscope in the 16th century. Although fungal spores were first observed by Giambattista della Porta in 1588, the seminal work in the development of mycology is considered to be the publication of Pier Antonio Micheli's 1729 work *Nova plantarum genera*. Micheli not only observed spores, but showed that under the proper conditions, they could be induced into growing into the same species of fungi from which they originated. Extending the use of the binomial system of nomenclature introduced by Carl Linnaeus in his *Species plantarum* (1753), the Dutch Christian Hendrik Persoon (1761–1836) established the first classification of mushrooms with such skill so as to be considered a founder of modern mycology.

Later, Elias Magnus Fries (1794–1878) further elaborated the classification of fungi, using spore colour and various microscopic characteristics, methods still used by taxonomists today. Other notable early contributors to mycology in the 17th–19th and early 20th centuries include Miles Joseph Berkeley, August Carl Joseph Corda, Anton de Bary, the brothers Louis Rene and Charles Tulasne, Arthur H. R. Buller, Curtis G. Lloyd, and Pier Andrea Saccardo. The 20th century has seen a modernization of mycology that has come from advances in biochemistry, genetics, molecular biology, and biotechnology. The use of DNA sequencing technologies and phylogenetic analysis has provided new insights into fungal relationships and biodiversity, and has challenged traditional morphology-based groupings in fungal taxonomy.

Fungal Endophytes: Common Host Plant Symbionts but Uncommon Mutualists

Endophytic fungi are an important, yet relatively unstudied group of microbial plant symbionts. Endophytic fungi live asymptomatically, and sometimes systemically, within plant tissues (Carroll, 1988, 1991). Endophytes usually inhabit above-ground plant tissues (leaves, stems, bark, petioles and reproductive structures), which distinguishes them from better known mycorrhizal symbionts. The distinction is not firm, because endophytes may also inhabit root tissues. Overall, endophytic fungi are ubiquitous and extremely diverse in host plants. Every plant examined to date harbours at least one species of endophytic fungus and many plants, especially woody plants, may contain literally hundreds or thousands of species.

Like mycorrhizae, endophytic fungi are thought to interact mutualistically with their host plants mainly by increasing host resistance to herbivores and have been termed "acquired plant defenses". Indeed, some agronomic grass species infected with systemic endophytes show striking toxic and noxious effects on vertebrate and invertebrate herbivores and pathogens (e.g., Gwinn and Gavin, 1992) by virtue of alkaloids such as pyrrolizidine alkaloids, ergot alkaloids and peramine produced by the fungi.

Endophytes, at least systemic ones in agronomic grasses, may also increase host grass competitive abilities, by increasing germination success, resistance to drought and water stress and resistance to seed predators . In return, plants provide spatial structure and protection from desiccation, nutrients and photosynthate and, in the case of vertical-transmission, dissemination to the next generation of hosts.

However, more recent arguments and evidence suggest that interactions between host plants and endophytes are not fixed in either ecological or evolutionary time, or geographically and range from mutualistic to antagonistic. This view is in keeping with more recent and general concepts of species interactions, and mutualisms in particular. For example, many plant-mycorrhizal interactions, the belowground counterparts of endophytes, are now recognized as ranging from mutualistic to antagonistic, depending on phylogeny, genetic strains, other interacting species, geography and abiotic conditions.

While systemic endophytes in agronomic grasses have been well-studied, the interactions between host plants and endophytes in natural populations and communities are poorly understood. The emerging picture from the limited studies of horizontally (spore) transmitted endophytes in plants suggests they:

1) are very abundant and common as localized infections in all types of plants, ranging from algae to angiosperms,

2) are extremely diverse, particularly in the more structurally complex and longer-lived woody plants,

3) have the same attributes of other macro-communities, including seasonality, successional changes, dominant and rare species, and generalist and specialist species , and

4) do not appear to generally increase host plant resistance to herbivores , as originally hypothesized.

Instead, many of these horizontally transmitted endophytes do not affect, or may even decrease, resistance to host plant herbivores. Herbivore damage to host plants may facilitate colonization of the spores and hyphae by breaching leaf surfaces and spores and hyphae may be dispersed via passage through the gut of herbivores (Craven, Wilson and Faeth, unpublished data). One thus does not expect that these endophytes should deter or decrease survival of herbivores.

Alternatively, systemic grass endophytes, at least in some introduced agronomic grasses, as well as a few native grasses may have profound effects on herbivores. Epichloe and Neotyphodium endophytes in these introduced grasses cause toxicosis to grazing livestock , and increase resistance to invertebrate herbivores and pathogenic microorganisms and their natural enemies and may inhibit germination and growth of other grasses via allelopathy by endophyte alkaloids. Neotyphodium-linked alkaloids (ergot and indole diterpenetype alkaloids) produce "staggers" (a neurological disorder) in sheep and cattle, while in tall fescue, pyrrolizidine and ergot-type alkaloids cause gangrene of extremities, reduced conception, and generally poor health in livestock. Resistance to insect pests in infected tall fescue and perennial ryegrass is mainly the result of peramine and pyrrolizidine alkaloids in tall fescue and ryegrass. While endophytes may confer other benefits to their hosts, such as increased drought resistance , alkaloids produced by symbiotic endophytes mediated many known benefits.

However, the beneficial effects of endophytes, especially those related to herbivory, are much less clear in native grasses. For example, Neotyphodium infections in most native grasses are not toxic to livestock and other vertebrates or invertebrates. Endophyte frequencies also tend to be more variable than agronomic grasses, although often high in some populations. Neotyphodiumlinked alkaloids, the main mechanism for endophyte— related benefits for the host grass, are also more variable in native grasses. Generally, alkaloid types tend be fewer and levels of individual alkaloids lower and more variable than agronomic grasses Variation in alkaloid levels and types in native grasses are probably linked to:

1) increased genetic heterogeneity of both host grass and endophyte,

2) more variable environments and

3) the interaction between variable genotypes and environments.

In general, there are relatively few cases of strong effects of systemic endophytes in native grasses on native herbivores. Even well-known cases of native toxic grasses appear limited to restricted few populations.

We propose here that variable frequencies and toxicity of systemic endophytic infections in natural populations may be explained by relative costs and benefits of harbouring the endophyte and their associated alkaloids. We first show that alkaloid levels are highly variable for Neotyphodium-infected plants within and among populations of Arizona fescue. We then develop a graphical model using nitrogen as a common denominator for costs and benefits of harbouring endophytes.

From this model, we make predictions when uninfected and low and high alkaloid-producing endophytes should persist in populations. Finally, we describe an experiment testing the effects of nutrients on growth rate of four grass genotypes that harbour Neotyphodium infections, and the same genotypes from which the endophyte has been experimentally removed.

Methods

Study System

Festuca arizonica (Vasey), Arizona fescue, in the subfamily Pooidaceae, is a dense, perennial bunch grass that reproduces primarily by seed (USDA, 1988) and is native and widespread in the Southwest. Arizona fescue occurs in semi-arid ponderosa pine (Pinus ponderosa)/ bunchgrass communities above 2,000 m elevation. Neotyphodium is common in Arizona fescue populations in Arizona. At least two varieties have been described from Arizona fescue based upon variable spore size and colour that appear in culture and molecular DNA and microsatellite DNA haplotypes.

Alkaloids

Alkaloids were determined from samples of infected plants from two Arizona fescue populations, Merritt Draw and Buck Spring, which are separate drainages about 3 km apart. Infected plants produce only peramine, often at low levels. Peramine concentrations were determined by L. P. Bush at University of Kentucky. Methods for peramine determination can be found in Bush et al. (1997) and Leuchtmann et al. (2000).

Model Development and Parameters

We developed a graphical model using nitrogen as the common currency for plant growth, endophyte growth and alkaloid production and herbivore consumption of the host plant. This approach seems reasonable because available nitrogen, especially in Arizona soils , is very low, and often limits plant growth. Furthermore, the primary basis for herbivore resistance via endophyte symbionts in grasses is fungal production of alklaloids, which are nitrogen-rich compounds. Alkaloids are well-known as plant-based defences but are costly to produce and may compete with other metabolic processes in plants which produce them (e.g., Ohnmeiss and Baldwin, 1994).

Herbivores, alternatively, are deterred by alkaloids in plants, but often respond positively to and increase consumption of plant tissues with increased nitrogen levels (e.g., Slansky and Rodriquez, 1982).

Finally, endophytes may alter nitrogen metabolism (e.g., Lyons et al., 1990) and increase nutrient uptake of host grasses by altering fine root structure and changing chemical environments near root zones. Alternatively, endophytes that produce alkaloids also may compete with the plant for nutrients, much like constitutively-based alkaloidal defences. However, endophytes vary in their capacity to produce alkaloid types and amounts, both genetically and environmentally. Thus, nitrogen demand appears as a common thread among endophytes, host grasses and herbivore, and our model is based upon nitrogen flux from the perspective of the host.

We model nitrogen flux in three hypothetical host grasses-an uninfected host grass, one infected with a systemic seed borne endophyte such as Neotyphodium, which produces low alkaloids and an infected grass with an endophyte that produces high alkaloid levels.

The graphical model was based upon the following assumptions: 1) presence of systemic endophytes can alter fine root structure and local soil environments (e.g., release of phenolic acids) such that nutrient uptake is enhanced at low soil nitrogen, 2) the endophyte uses nitrogen for its own growth and production of nitrogen-based alkaloids, 3) herbivores reduce plant nitrogen through consumption and, 4) the magnitude of 2) and 3) depend on the amount of alkaloids produced by the endophyte. Finally, we assume that 1-4 are functions of soil nitrogen content.

Experimental Test

Four infected 'mother' plants from the Merritt Draw study site were split into ramets and treated hydroponically with low levels of the fungicide propiconazole to remove the endophyte. Treatment removes Neotyphodium from about 50% of ramets (hereafter E-); the remaining infected ramets served as infected controls (hereafter, E F). Other ramets were treated hydroponically but without fungicide and thus remain infected (hereafter E).

The four mother genotypes were selected based upon their peramine alkaloid content, such that two genotypes were low in peramine (1.8 ppm). After hydroponic treatment, all ramets were planted individually into 16 oz. cups with native soil, and continually split and re-potted as they grew to provide cloned replicates (at least 5 of each).

After one year in the greenhouse, these plants were transplanted into a plot at the Arboretum of Flagstaff (1 m apart) in spring 1998 and randomly assigned the following treatments: 1) ambient water and ambient nutrients 2) supplemented water (1,000 ml per plant every 3 days), 3) added nutrients (1.5 g/liter- 30-15— 30 fertilizer, every two weeks and 4) added water (every 3 days) plus supplemented nutrients (every 2 wk). Ambient soil nutrients are very low (e.g., 1-2 ppm N) in native soils at the Arboretum (Schulthess and Faeth, 1998), and this level of fertilizer significantly increases nitrogen content of grass plants (Saikkonen et al., 1999). Likewise, ambient precipitation at the Arboretum is typically zero from the end of summer rains in late August to beginning of winter precipitation in November and December and very low during May-June. The plot (except for area near the plants) was covered with a weed barrier (Dalen Co.) that permits water and nutrient penetration but prevents weed growth and hence unwanted plant competition.

The Arboretum is fenced (4 m high) to exclude livestock and native grazers (elk and deer). Thus, this experiment was conducted under conditions of low herbivory; only small rodents and invertebrates have access to plots.

We measured the rate of growth (diameter and height to estimate plant volume) at the beginning and end of the first growing season. All plants were tested via tissue print immunoassay (Schulthess and Faeth, 1998) to confirm endophyte status. We used a factorial ANOVA to compare differences in mean rate of growth (change in volume within the first growing season). All assumptions of ANOVA were tested and met.

Results

Graphical Model

To build our graphical model of plant nitrogen flux, we first assume that endophytes stimulate nitrogen uptake according to a decaying exponential function as soil nitrogen increases. We only consider environmental conditions under which plant survival is possible. Here, "zero nitrogen" represents the minimum for survival, rather than zero soil nitrogen per se. Uninfected host grasses do not stimulate nitrogen uptake.

For simplicity, we assume that infected plants exist in two states, those with endophytes that produce high levels of alkaloids and those that produce low levels. This assumption is supported by preliminary results on peramine levels in natural populations of Neotyphodium-infected Arizona fescue although infected plants in natural populations produce a wider range of peramine than simply low or high.

Infected plants ranged from zero levels of peramine to 2.3 ppm peramine, levels high enough to increase resistance to invertebrate herbivores. We model these two types of infected plants by assuming that high alkaloid endophytes (HAE) use most or all of the nitrogen whose uptake they stimulate, plus additional nitrogen whose rate increases linearly with increasing soil nitrogen. Alternatively, low alkaloid endophytes (LAE) use nitrogen at a constant rate. For uninfected plants (No E), the curve = 0 (not shown).

Next, we assume herbivory removes nitrogen from the host plant via consumption at rates that are either independent of , or that depend on, soil nitrogen. The first case may represent generalist herbivores such as large grazing ungulates that do not adjust consumption rate based upon plant nitrogen content, but do respond to toxic or noxious allelochemicals.

In contrast, the soil nitrogen-dependent case may represent more specialized herbivores that increase consumption based upon soil nitrogen and thus nitrogen content of the host plant. In both cases, herbivores consume at a higher rate from LAE plants HAE plants. As suggested by past studies of Arizona fescue , herbivory on uninfected plants (no E) is basically the same as that on LAE plants.

Model Outcomes

Summation of the underlying terms, stimulation and endophyte use and losses to herbivory predicts net nitrogen flux in the host grass and suggests when different endophyte strains should be favoured.

At low soil nitrogen, LAE plants should increase net nitrogen flux in their host plants more than HAE plants. Because they do not benefit from endophytic stimulation, no E plants should exhibit a net nitrogen loss through herbivory and thus be at disadvantage, relative to LAE and HAE, at low soil nitrogen.

In contrast, at low soil nitrogen, both low and high alkaloid endophytes may produce a net positive nitrogen flux in their hosts. As least in terms of our nitrogen flux model, it is at low soil nitrogen levels that the interaction between host plants and endophytes are potentially mutualistic.

As soil nitrogen increases, however, all three host types (HAE, LAE, no E) show a net nitrogen loss. Under high soil nitrogen, No E hosts should be favoured, followed by HAE and then LAE. Where on the soil nitrogen axis that the shift in advantage from LAE to no E plants occurs depends critically on the relative magnitudes of the stimulation, usage, and herbivory terms.

Experimental Results

Presence of the endophyte increased growth rate, as estimated by change in above-ground volume during the growing season. However, plant genotype also strongly influenced growth rate.

Furthermore, growth rate varied differently across plant genotypes depending on the presence or absence of the endophyte (significant genotype X endophyte interaction, with genotypes 3 and 4 showing the largest positive response in growth rate when the endophyte was present. No other effects or interactions were significant; however we present the endophyte X treatment interaction for sake.

Fungicide-treated but infected plants (fungicide controls E F) were not different in growth rate than fungicide-treated and endophyte-removed (E-) plants (F = 2.81; df = 1, 19, P = 0.11), indicating no extraneous effect of fungicide on growth of grasses. Likewise, treatment and endophyte status (E- and E F) did not interact significantly (F = 1.65, df = 3,19, P = 0.21).

Discussion

The simple graphical model suggests that the relative costs and benefits of harbouring endophytes, at least in terms of nitrogen flux, change along soil nitrogen gradients. Furthermore, from the perspective of the host plant, relative benefits and costs of harbouring endophytes depend on whether the endophyte produces high or low levels of alkaloids and the intensity of herbivory along these gradients.

The curve intersections, also indicate a shift in advantage from grass hosts that contain symbiotic endophytes to host grasses that do not. The model further suggests that low alkaloid-producing endophytes have greater relative benefits at the lowest levels of soil nitrogen, with the magnitude of this benefit diminishing as soil nitrogen increases.

The parameters of these graphical models have been roughly estimated and more precise measures of endophyte-induced uptake, endophyte usage, and herbivore consumption may change the relationships, especially the location of intersection points for net nitrogen flux. Intersection points are of particular interest because they represent predicted shifts among host-endophyte types along a soil nitrogen gradient.

Empirical data on model parameters are needed to pinpoint where, and how often, these intersections occur. Nevertheless, the graphical model serves as heuristic framework for how endophyte-host interactions may change in varying environments. This is especially important because almost all experimental studies of systemic endophytes in grasses have used the introduced grasses, tall fescue and perennial ryegrass, in agronomic settings, where nutrients are typically supplemented and herbivory is more chronic than in natural settings. For many native grasses, such as Arizona fescue, which are found in low nutrient soils and experience sporadic herbivory , the interactions among endophyte-types and host grasses may be occurring to the left or near the intersection points. If so we may expect a dynamic shifts in fitness advantages among uninfected and low alkaloid and high alkaloid endophyte-infected plants with relatively small changes in herbivory and available nutrients.

The models also indicate that simply considering whether a host plant is infected or uninfected with an endophyte is not adequate, because costs and benefits to host plants depend on how much alkaloid the endophytic symbiont produces. Most previous experimental studies of the effect of systemic endophytes on host performance or resistance to herbivores have used varieties of infected tall fescue or perennial ryegrass grass compared to an uninfected counterpart. These varieties often have less host and endophyte genetic variation and therefore less variation in alkaloids than native grasses.

Recent studies have identified genes responsible for alkaloid production, but abiotic environments can also modify expression (e.g., Wilkinson et al., 2000). For Arizona fescue, it is clear that even within a natural population, infection by Neotyphodium can result in a wide range of peramine levels. This variation appears mostly determined by endophyte haplotype but also modified by host genotype and local environments.

For Arizona fescue, as well as other native grasses harbouring systemic endophytes (e.g., Faeth, 2002), most natural populations are mosaics of uninfected and infected host plants, the latter of which greatly vary in alkaloid levels. These mosaics may be maintained by shifting costs and benefits as suggested by our model.

Our experimental results partially support the possibilities outlined by the graphical model. Notably, the presence of the endophyte increases growth rate overall, suggesting an overall advantage to harbouring the endophyte in Arizona fescue, at least in terms of growth rate and under restricted herbivory (large grazing vertebrates were excluded). Furthermore, although the interaction of endophyte and treatment is not significant, the mean growth rate of infected plants increased in all treatments except the high water and high nutrient treatment, suggesting that advantages of infection diminish with increased soil moisture and nutrients. These results corroborate those involving agronomic tall fescue, where host performance of infected plants diminished at high soil nutrient levels (e.g., Cheplick et al., 1989).

More notable, however, is the strong effect of plant genotype on growth rate and the interaction of plant genotype with the presence or absence of the endophyte. Plant genotype 3 and especially 4 (genotypes 3 and 4 are significantly different from 1 and 2, P 1.8 ppm peramine) while genotypes 1 and 2 produce either zero or only trace amounts.

Both the model and experiment suggest that the direction and strength of endophyte-host plant interactions depend on plant and endophyte genotype, and environmental conditions. The model included only nitrogen as the common currency, but clearly water availability also influences growth response of the host-endophyte symbioses. Nevertheless, because many of the purported benefits, as well as the costs, of symbioses between endophytes and host grasses are related to nitrogen budgets, this may be an good starting point for explaining differences in endophyte frequencies among and within grass populations.

Conclusions

Our model and experimental results suggest an explanation for the observed variation in Neotyphodium infection frequency and in alkaloid production by infected grasses in natural populations. Under varying herbivory and soil nutrients and moisture, cost and benefits of harbouring the symbiotic endophyte shifts. Furthermore, all infected grasses are not equivalent but depend on host and endophyte genotype, which in turn influences alkaloid production and its associated cost and benefits. This view of variation in endophyte-host plant symbioses is necessarily more complex than previous ones involving endophyte-agronomic grass symbioses, but is more reflective of endophyte and grass interaction in natural populations and communities.

2

Microbiology in Biotechnology

Microbiology in Biotechnology

Activities of microorganisms are very important to almost every sector of concern to mankind. From a perusal of the foregoing topics, one can find applications (uses) of microorganisms to agriculture, forestry, food, industry, medicine, and environment. The scope and significance of microbiology has enlarged manifold, particularly when importance of environment was realised globally and the word environment was used in a much wider sense in terms of totality to include almost everything, every bit of nature. Demographic pressure on land with subsequent rise in demand for food and shelter had, out of several, two apparent impacts on nature and its resources -

(i) contraction (leading to deplection and in some cases finally to extinction) of biodiversity (fauna, flora and microbes), and

(ii) degradation of environment. Land became denuded, cultivable land area squeezed, air and water turned to be unfit for man, and wastes from urban, municipal and industrial areas piled up all around. Natural resources started depleting and there developed an energy crisis. It is in this context that the role of microbes was realised in order to exploit their vast potential to possible solution of some such problems.

New strains of microbes have been developed having the desired activities (entirely a new activity or enhancement activity in terms of rate productions). Environmental microbiology, a new field, could thus become of much relevance to us.

Most of these activities of microbes to agriculture, forestry, industry, medicine, food and environment have already been referred. In this chapter an overview of this subject has been presented with the simple purpose of just a quick revision to remind the importance of microorganisms to human race

Microbes and Agriculture

Microbes and Agriculture - Besides being important in biogeochemical cycling of nutrients, microbes play vital role in maintenance of soil fertility and in crop protection.

Soil Fertility

Soil Fertility - Microbes are being exploited in two important ways - biofertilisers, and creating new nitrogen-fixing organisms.

Biofertilisers

Potential of Rhizobium, Azotobacter, Beijerinckia, Azospirillum, Cyanobacteria, such as species of Aulosira, Anabaena, Nostoc, Plectonema, Scytonema, Tolypothrix, and Azalia as biofertilisers has been exploited so as these could serve an alternative to chemical fertilisers.

Many brands of rhizobial inoculants are already in market today in the country. Several organisation and manufacturers are producing huge quantities of Rhizobium culture in the country.

These include Micro Bac., India, Shyam nagar, Parganas; Bacifil Inoculants, Lucknow; Govt. of Tamil Nadu; Nitro Fix Industries, Calcutta (W. Bengal) and Indian Organic Chemical Ltd. Bombay. In some other States, units are being prepared for increase in its production. Much progress has also been made with cyanobacteria in this direction.

Mycorrhizae, both ecto and endomycorrhiza help in uptake of N, P, K and Ca. They, particularly help in phosphorous nutrition.

Nitrogen Fixers

Nitrogen Fixers - Through recombinant DNA technology efforts have been made to introduce nitrogen-fixing genes (nif genes) into wheat, corn, rice, etc. Plasmids of the bacterium, E. coli and yeast are being worked out for such a possibility. Hybrid E. coli plasmid cloned with nif genes of a nitrogen-fixing bacterium, Klebsiella pneumoniae and hybrid yeast plasmids are then integrated.

Biopesticides

Biopesticides - Several microbes (viruses bacteria, and fungi) are being developed as suitable biopesticides for management of insect and nematodal pests. Some fungi have good potential of their use as bionematicides to control nema-todal pests of vegetables, fruit and cereal crops. Some bacterial and fungal products are also in use to control diseases of roots and shoots of plants.

Bioweedicides

Bioweedicides - Several fungi have been found very useful in the control of troublesome weeds of crop fields. Registered products are available in market for use in several countries.

Microbes And Public Health

Microbes And Public Health - The most successful bioinsecticide has so far been the bacterium, Bacillus thuringiensis. Several registered products of different strains of this microbe are on sale, thuricide being one, for the control of insects including mosquito-the carrier of malaria.

Microbes And Food

Microbes And Food- Applications of microbes to food, energy, industry and environment are interrelated in most situations. To cite an example, degradation of urban, municipal and industrial wastes by using a suitable (may be a tailored, genetically-engineered) strain of a microorganism should result into (i) dis-posal of pollutant, (ii) biotransformation of a waste- into a byproduct, suitable for consumption as food (conversion of agricultural waste into single cell protein is an example) and (iii) production of energy during such conversions. Therefore it is a three-way beneficial process carried out by a microbe.

Several yeasts, Saccharomyces, Candida, Torulopsis can be grown on waste materials, recycling them into food. Crude petroleum products are converted to SCP by Methylophilus methylotrophus. Cattle dung is fermented by methanogenic bacteria to biogas, an ideal biofuel in rural areas. Mushroom cultivation is good example of SCP where agricultural waste is recycled into food (SCP) and the left over residue used as organic manure.

Microbes And Industry

Microbes And Industry - Applications of microbes in industry are well known. Various microorganisms are used for commercial production of alcohols, acids, fermented foods, vitamins, medicines, enzymes etc. One recent development in industrial microbiology has been the production of immobilised enzymes and cells for production of these chemicals at enhanced rates with simultaneous recovery of the enzyme(s) involved in such processes.

New strains of microbes have also been developed through recombinant DNA technology for overproduction of metabolites. Immobilised enzymes and cells could have their maximum application in industrial microbiology. Immobilised enzymes have also been utilised in medicine.

Acetone Butanol Fermentation

Acetone Butanol Fermentations -The acetone butanol fermentation is one of the oldest fermentation known. The fermentation is based on culturing various strains of Clostridia in carbohydrate rich media under anaerobic conditions to yield butanol and acetone.

Clostridium acetobutylicum is the organism of choice in the production of these organic solvents. These fermentations were out of favour till very recently because of the availability of acetone and butanol from the petroleum industry.

Today there is considerable amount of interest in these fermentations. However, the concentration of end products in these fermentations is quite small and the fermentations are a type of mixed fermentation yielding a mixture of compounds such as butyric acid, butanol, acetone etc. Attempts to increase yeilds by use of genetidally altered strains or change in fermentation conditions have been partially successful ocassional careers, carried out in a professional way.

Microbes In Recovery of Metals

Microbes In Recovery of Metals - Recently microbes have been found very useful in enhanced recovery of metals including uranium from low-grade ores. Through bioteaching these microbes are able to solubilise the metals from their ores. Microbes thus play important role in mining and recovery of metals. For instance, Thio-bacillus thiooxidans and T. ferrooxidans can be used in recovery of copper.

Biotechnology in Agriculture : The New Green Revolution

The New Green Revolution

The green revolution which gave us plenty of grains to feed millions of people and the revolution in medicine which increased the life span of man are common knowledge even to lay people. All this was possible due to major discoveries and technological innovations in agriculture and medicine. Today, we are witnessing another revolution in biosciences be-cause of some major advances in cell biology and genetics.

Many people are inclined to believe that while the battle for green revolution of the type we saw three decades ago was fought in the field, the battle for the new revolution in biosciences known as "Biotechnology" is being fought in modem laboratories.

Some argue that the new biotechnology in agriculture is the second green revolution (part II) which will speed up crop improve-ment by gene manipulation in a Petri dish rather than in an open field.

The meaning or the definition of the word biotechnology has been the subject of hot debate by scientists and technocrats. The definition depends on the extent of expertise a group or an individual has with regard to cell biology. It also depends on the needs of a society or a country one lives in. The use of microorganisms or their products for food, feed, biofertilizers, biopesticides and medicine was known during the last 60 years.

The major developments in medicine and industry during those years came with the use of microorganisms for the benefit of mankind through new inventions in microbial technology and fermentation processes but the biotechnology of which we are currently talking about is envisaged by splicing genes and altering genomes by insertion of foreign DNA by genetic recombinant DNA techniques (the so-called genetic engineering).

The cell which is being manipulated by genetic recombination may be a microbial cell or a cell from a tissue of a plant, animal or man with the ultimate objective of inserting or cloning useful genes to obviate the use of long and tedious process of conventional breeding often replacing it by tissue culture techniques.

David Baltimore, Nobel Laureate, formerly Director, Whitehead Institute and Professor, Department of Biology, Mas-sachusetts Institute of Technology defined Biotechnology as the applica-tion of scientific and engineering principles to the processing of materials by biological agents to provide goods and services. The author defines biotechnology as the application of science and technology to accelerate or improve nature's processes in producing man's ever increasing needs for a good living. The path from green revolution to gene revolution or from conventional plant breeding to genetic engineering has been filled with many significant findings. For almost a century plant breeders identified and selected desirable characters and combined them into one individual plant. Since all characters are controlled by genes in chromosomes, plant breeding may be regarded as manipulation of chromosomes.

This was done by the sorting and retention of similar chromosomes in the same plant to reach a homozygous state, a method termed pure line selection. Alternatively, different chromosomes can be combined to form a heterozygous state, a method known as hybridization conferring hybrid vigour or heterosis.

The next step was the development of genetic variability through spontaneous or artificially induced mutations. Normal plants are diploids but when plants are developed with three or more sets of chromosomes, they become polyploids that' tend to be bigger than diploids. Autopolyploid plants have genes similar to their diploid ancestors whereas allopolyploids are combinations of genomes of two different species that differ in characteristics.

The first achievement of hybridization techniques was the development of hybrid maize in 1919 which revolutionizep American agriculture. The development of hybrid wheat and rice plants in 1960s filled the bread basket of developing countries, generally known as green revolution, for which Dr. Norman Borlaug was awarded the Noble Peace Prize in 1970.

The discovery that plant cells can develop into entire plants was another land mark in the development of new varieties of crop plants. The term tissue culture was coined to denote the in vitro development of plants in test tubes from calluses generated from plant parts. This led to mass production of uniform plants and revolutionized floriculture in the globe. Tissue culturing often leads to progenies which are variable. These progenies are known as somoclonal variations and have been exploited to generate mutations and it has been estimated that in vitro tissue cultures can produce ten times more somoclonal variations than that can be induced by chemical mutagens.

Fusion of naked genetically compatible protoplasts (inter specific) resulted in successful regeneration of new varieties. Fusions between incompatible protoplasts (inter generic) resulted in abortive cell division and successes in regeneration was never achieved, excepting the instance of crossing between tomato and potatoes forming 'pomatoes' which can only be regarded as a laboratory success not amenable to commercial exploitation. With the advent of biotechnology, agriculture has reached a science based industrial state. By using recombinant DNA technology, many transgenic life forms have been engineered since 1985. Transgenic plants belonging to both monocotyledonous plants such as maize, millet, wheat, rice and ragi and dicotyledonous plants such as alfalfa, clover, peas, soybean, mothbean, potato, tobacco, cotton, flax, sugarbeet and sunflower have been constructed. New varieties of vegetables and fruits such as cabbage, carrot, cauliflower, celery, cucumber, horseradish, lettuce, rape, grape, muskmelon and strawberry have been developed. The new varieties have incorporated genes capable of resisting one or more of the following: herbicides, insects, stress, frost or virus infections.

Plant biotechnology has opened up the possibility of producing artificial seeds, artificial sweetners (sugar substitutes) and bioplastics. Normal seeds have an embryo surrounded by cotyledons for initial sustenance during germination. By somatic hybridization, plant embryos can be mass multiplied in fermentation tanks and each embryo is then encapsulated in a jelly-like coat that can be called an artificial seed. Some estimates have revealed the possibility of production of 80,000 embryos per day but the cost could well be prohibitive for commercial exploitation. Presently, several companies are engaged in reducing the cost for atleast some crops such as carrots and celery.

The most important sugar substitute is the maize based high fructose com syrup (HFCS) known as isoglucose in Europe. Some estimates put HFCS production worldwide to 6 million tonnes available in liquid as well as crystal form.

Aspartame is a synthetic chemical thousand times sweeter than sugar. With the advent of this product nearly 38 research institutes and companies around the world are engaged in producing novel chemical sweetners.

Thaumatococcus danielli or commonly known as Katemfe is a plant that grows in humid forests in Western and Central Africa. The berries of this plant contain the protein thaumatin that is 2500 times sweeter than sugar. Tate and lyle, a UK based sugar company had set up plantations of Katemfe in Ghana, Liberia and Malaysia. The frozen berries were processed to obtain purified thaumatin that was sold under the brand name 'Talin'. One drawback of thaumatin is its lingering taste limiting its use in food products. In spite of this, research is underway to understand the gene coding for thaumatin and its transfer to E. coli and other plants.

Stevia rebaudiana grows in Paraguay and several countries in South East Asia. The plant is capable of producing proteins several hundred times sweeter than sugar. The product is being marketed in Japan which has also bitter taste. African forests abound in sweet berries such as 'Miraculous berry' (3000 times sweeter than sugar) and Mexico has Lippia dulcis, thousand times sweeter than sugar. The search for cheaper substitute to cane sugar is being pursued vigorously and in future years we may have alternate sugar sources.

An interesting example of how a plant can be made to produce novel chemicals such as bioplastics is the transgenic Arabidopsis capable of producing granules of polyhydroxybutyrate (PHB), a polyester which is normally obtained from the bacterium Acaligenes eutrophus. In fact, PHB is a storage product in many bacteria intended to be used as a source of energy by bacterial cells in times of nutritional stress.

This bioplastic material is a delicate product destroyed by pH above 8 and temperatures above 70°C. The product is mixed with polyhydroxyvalerate (PHV) to make it flexible and moulded into any shape, spun as fibre or rendered into a film. It is biodegradable to CO_2 and H_2O with no environmental hazard. The bioplastic is compatible with living tissue and hence can be adapted for medical purposes. It can also be used as a much in agriculture.

Transgenic animals and microorganisms are being used for fundamental research, for production of pharmaceuticals such as goats-Iactoferrin and for production of biological control agents. Pigs and rabbits are genetically engineered to function as organ donors for human beings and chickens have been exploited for producing foreign proteins in eggs. Pharm biotechnology (agricultural production of pharmaceutical products) has to be distinguished from Farm biotechnology (productivity related agricultural applications) and very likely agriculturally produced pharmaceuticals will be marketed sooner than agricultural products, despite the fact that the latter could undoubtedly increase global food supply.

This has been due to success in pharmaceutically oriented animal experiments as exemplified by the transgenic modification of pigs and sheep for expressing valuable pharmaceutical products in milk where all animals in the offspring appeared healthy contrasting with the transgenic pigs generated to produce leaner meat or more rapid growth whose offspring had adverse effects.

There are about 20 different man made pharmaceuticals involving crops that have been genetically changed to produce a range of prophylactics from cholera vaccines, herpes vaccine and cancer treatments. Potatoes seem to be ideal vehicles for the new generation of vaccines such as vaccine against E. coli disorders of the intestine. These are friendly and easier to tolerate than injections.

Phosphorus in seeds is a poor nutrient for monogastric animals such as chickens unless phytase is present to release phosphorus. Feeding chickens with seeds containing the phytase gene from Asperigillus niger brought about growth increases in chicken. This biotechnological innovation known as "gene farming" not only improved the quality of chicken feed but also minimised the excretion of phosphate in the environment. Another example is the case of sweet potato which is a staple food in China. The strategy here was to implant twin genes such as viral coat protein gene and Bacillus thuringiensis genes into sweet potato to ward off diseases caused by viruses as well as insect pathogens.

Nucleic Acids

It would be helpful to briefly describe some basics of molecular biology before attempting to understand its implications in biotechnology. The classical discovery of the structure of genetic material by Watson and Crick in 1953 revealed the unique suitability of nucleic acids for carrying genetic information and transmitting to subsequent generations.

All the information needed for growth and multiplication of most organisms is carried by nucleic acids, especially the double-stranded deoxyribonucleic acid (DNA) or single or double stranded ribonucleic acid (RNA).

RNA differs from DNA in that the single strands have a ribose instead of deoxyribose and uracil in place of thymine. The double stranded DNA occurs in a helical structure with a backbone of alternating phosphate and deoxyribose molecules having a purine or a pyrimidine base linked to the I-position of each sugar molecule. The two complementary strands of DNA are twined together by hydrogen bonds between the purine and pyrimidine base pairs: adenine-thymine (AT) and guanine and cytosine (CT). An adenine in one strand of DNA occurs directly across a thymine in the other strand. Similarly, a guanine (G) occurring in one strand is bonded to a cytosine (q across the other strand).

The genetic information is coded by the linear arrangement of bases on the DNA strands. The sequence of nucleotidase dictates all the characteristics of an organism and serves as a genetic code. The sequence of nulceotides, read in groups of three (triplet) reflects the sequence of amino acids in the large number of proteins (enzymes) synthesised by a cell. Each triplet is known as a codon and there are 64 possible combinations beginning with four nucleotides.

Protein Synthesis

Mitosis (cell division) results in the formation of tissues. Before cell division takes place, the chromosomes in the parent cell have to be duplicated so as to be equally shared by the daughter cells. This replication is carried out within the nucleus of the cell by the action of an enzyme known as DNA polymerase.

The synthesis of proteins takes place within the cytoplasm of the cell away from the nucleus. The genetic information in the DNA is conveyed to ribonucleic acid in ribosomes residing in the cytoplasm by a process known as transcription. During transcription, only one of the two strands of DNA becomes translated into RNA by RNA polymerase. The synthesis of RNA always proceeds in a fixed direction beginning at the 5' end and concluding with 3' ended nucleotide.

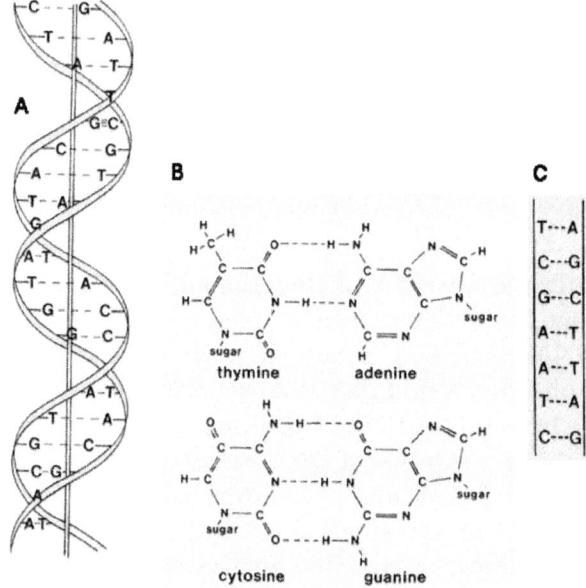

Figure 1: The double helix. a. Deoxyribonucleic acid (DNA) is a double-stranded, helical molecule composed of nucleotide units. b. Diagram of the molecular structure of the base pairs of DNA. c. Diagram showing pairing of the nucleotide bases in a short DNA segment.

The synthesized RNA from DNA of the nucleus in a cell moves into the ribosomes of the cytoplasm carrying with it information needed to synthesize protein in the ribosome. Hence, this RNA has come to be known as messenger RNA (mRNA). The 5' end of an mRNA molecule attaches to a ribosome.

		U		C		A		G	
U	UUU⌉ UUC⌋	Phe	UCU⌉ UCC		UAU⌉ UAC⌋	Tyr	UGU⌉ UGC⌋	Cys	U C
	UUA⌉ UUG⌋	Leu	UCA UCG⌋	Ser	UAA UAG	Stop Stop	UGA⌉ UGG⌋	Stop Trp	A G
C	CUU⌉ CUC		CCU⌉ CCC		CAU⌉ CAC⌋	His	CGU⌉ CGC	Arg	U C
	CUA CUG⌋	Leu	CCA CCG⌋	Pro	CAA⌉ CAG⌋	Gln	CGA CGG⌋		A G
A	AUU⌉ AUC	Ile	ACU⌉ ACC		AAU⌉ AAC⌋	Asn	AGU⌉ AGC⌋	Ser	U C
	AUA⌋ AUG	Met	ACA ACG⌋	Thr	AAA⌉ AAG⌋	Lys	AGA⌉ AGG⌋	Arg	A G
G	GUU⌉ GUC		GCU⌉ GCC		GAU⌉ GAC⌋	Asp	GGU⌉ GGC⌋	Gly	U C
	GUA GUG⌋	Val	GCA GCG⌋	Ala	GAA⌉ GAG⌋	Glu	GGA GGG⌋		A G

About 4 per cent of total cellular RNA is mRNA. Since only a small segment of mRNA is attached at a given time to a ribosome which is moving across it, a single mRNA molecule can be read at the same time by several ribosomes, occurring as polyribosomes consisting of anywhere from 6 to 50 ribosome units.

Before the process of amino acid polymerization into proteins begins, the 20 different amino acids in a cell are first transformed into energy-rich precursors. These precursors get attached to a small transfer RNA (tRNA) molecule. A group of three nucleotidase (triplet) constitute a tRNA and serves as an codon that uses base pairing to find three nearby nucleotidase which in turn serves as a codon on a mRNA molecule.

Specific enzymes known as aminoacyl synthetases now begin to bind or attach aminoacids to specific tRNA molecules. Ribosomes are thus minifactories for protein synthesis by a series of codon-anticodon interactions which occur on their surfaces.

These interactions take place when mRNA molecules move across the active surface of ribosomes aligning successive codons into position to form successive aminoacids along polypeptide chains and this process is known as translation.

There are 64 potential codons and of these 61 are used to specify amino acids while 3 are made use of to provide signals to terminate the formation of polypeptide chains. Several amino acids are determined by more than one codon. All the codons put together constitute the genetic code and this code is common to all forms of life including microorganisms.

Southern and Northern Blot Techniques

Southern blot is the classic technique described by Southern in 1975 for understanding individual genes in a complex mixture of DNA.

The pro-cedure involves cutting up high molecular weight genomic DNA into fragments by enzymatic digestion with restriction endonucleases. These bacterial enzymes have been isolated and are available in many laboratories.

They have been prepared from several bacteria and are designed as molecular scissors which cut the strands of DNA at specific oligonuleotide recognition sequences. The fragments of DNA are then separated on the basis of size by agarose gel electrophoresis. The fragments with smaller number of nucleotides move faster on the gel in an electric current than the larger ones which have higher number of nucleotides.

At the end of the experiment, the gel is immersed in a fluorescent dye (ethidium bromide) which binds to DNA. When viewed under ultra-violet light, the fragments can be visualized as separate entities reflecting the nucleic acid pattern. The cellular RNA can also be run on agarose gel and the fragments can be separated in much the same way as DNA. This technique is known as Northern blot technique.

Recombinant DNA Technology and Gene Cloning

Man made nucleotide sequences tailored to carry desired traits, known as molecular probes have been extensively used in molecular biology studies. These nucleotide sequences (probes) ought to be correct, pure and available in large amounts. The probe must be labelled in some manner to help detection. These probes are prepared by recombinant DNA techniques using the common enteric bacterium *Escherichia coli*. This organism has been studied so often by molecular biologists that they are now aware of the implications of its genome containing about 4.2 million base pairs (bp).

Furthermore, the bacterium has extrachromosomal DNA molecules known as plasmids that can replicate autonomously independent of the nuclear DNA. These plasmids are closed single pieces of supercoiled DNA which can be stably inherited by daughter cells. They have the ability to replicate to high numbers (copies) within each bacterial cell.

The restriction endonucleases cleave DNA only to specific oligonucleotide sequences rather asymetrically leaving 'sticky' ends. The sticky ends remain complementary between any two DNA fragments sliced by the same restriction enzyme. This property makes it easy to insert or 'clone' an outsider or foreign gene to the E. *coli* plasmid provided it has been sliced by the same restriction enzyme.

The transformed plasmids known as 'vectors' can be amplified to a high copy number in a standard bacterial culture. To retrieve the inserted foreign DNA, the bacterial biomass from the culture medium is separated and lysed. The DNA content is purified and sliced by the same restriction enzyme that was earlier used for original cloning.

Plasmid vectors have been used to transfer DNA from one prokaryotic cell to another, from a prokaryote to an eukaryote and from an eukaryote to a prokarotic cell. The plasmid vectors have the limitations of cloning upto 5000 base pairs (bp) or 5 kilobases (kb). By developing and using bacteriophage lambda chromosomes, foreign DNA have been cloned upto 15 kb.

The ability to use vectors has been further enlarged by using features of both plasmids and bacteriophage lambda to the extent of 50 kb. Presently, specially constructed DNA fragments from yeast cells known as yeast artificial chromosomes (YAKs) are available that can be used to clone pieces of DNA upto 1 million bp.

One of the important steps in cloning a foreign gene is to obtain the desired gene in the absolute pure condition. This can be done by a traditional method beginning with the purified protein which the gene produces. Required antibodies are raised which will recognize and precipitate the protein when added to a cell extract from a tissue where the protein is actually synthesized.

This results in the precipitation of newly made polypeptides which are being elongated on the polyribosomes. However, the precipitate contains unwanted ribosomes mixed with the mRNA templates required for producing the protein in question. When these mRNA templates are purified from the mixture, a sequence of nucleotides complementary to the gene of interest can be obtained. By using a retroviral enzyme reverse transcriptase, which synthesizes DNA from a RNA template, a cDNA sequence from the mRNA sequence can be obtained.

By the addition of DNA polymerase the second strand of DNA can be replicated which results in the formation of a double stranded DNA copy of the gene of interest. This elaborate procedure is known as the polysome precipitation.

In recent years, polysome precipitation method has been replaced by 'DNA libraries'. A cDNA library is made up of all the actively transcribed genes of a tissue inserted into a population of bacterial cells. The bulk of mRNA preparation is reverse transcribed and inserted into plasmids in one lot, with the objective that every possible cDNA sequence will be carried by atleast one bacterial cell in the culture. This cDNA library has to be sorted out for a particular cDNA sequence of interest.

The bacterial culture containing the cDNA is spread out on agar plates and filter blot technique is used with a labelled oligonucleotide probe (1D-40 bp) to select a colony containing the foreign gene of interest. Presently, many biotechnological companies have developed automated instruments that can rapidly synthesize oligonucleotides of any sequence.

Gene Exchange in Bacteria

Gene e change in bacteria takes place by processes known as transformation, transduction and conjugation. The addition of foreign DNA to actively growing bacterial culture results in chance entry of foreign DNA into cells by modification of bacterial cell envelope followed by the intake of DNA into the bacterial genome.

Transduction involves bacteriophages or bacterial viruses whose DNA enters the bacterial cell followed by the disintegration of the bacterial chromosome. The phage DNA multiplies in the cells, the cell walls undergo lysis releasing the phages which have multiplied in the mean time.

The phages which mediate this type of transduction are known as lytic or virulent phages. On the contrary when a temperate phage (non-virulent type) enters the bacterium, the phage DNA becomes attached to the bacterial chromosome and remains integrated with the bacterial genome for many generations.

This process is also known as lysogeny. At times, temperate phages may turn virulent leading to lysis and production of more bacteriophages.

The temperate phage, when freed from the cell may carry with it small pieces of DNA which upon delivery to a next host cell can add an additional character to the new cell's capabilities.

Contact between two cells is required for conjugation, one acting as a donor and the other a recipient. The donor possesses a fertility factor (F⁺) and the recipient has no such factor (F⁻) but must be viable for successful conjugation.

The Agrobacterium Mediated Transfer of Genes

A naturally occurring conjugation phenomenon in *Agrobacterium tumefaciens* induced crown-gall disease of plants has been ably exploited by biotechnologists to genetically engineer foreign genes into dicotyledonous plants rendering them transgenic in characters such as resistance to viral diseases, herbicides or bioinsecticides.

A. Bacterial transduction with bacteriophages. A virulent phage

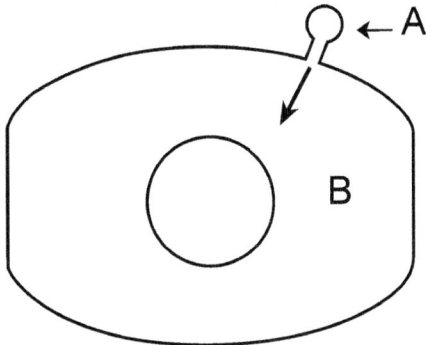

C. *The phage DNA is injected into the bacterial cell followed by disruption of bacterial DNA and the phage takes Control of bacterial cell functions*

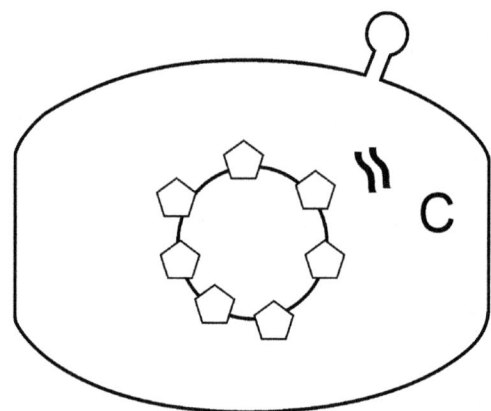

D. *New bacteriophages are formed and released by cell lysis*

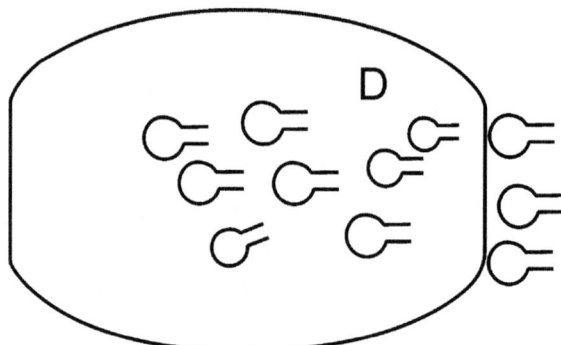

E. *In a phenomenon called lysogeny infection with a temperate phage results in the integration of phage DNA with the bacterial Chromosome*

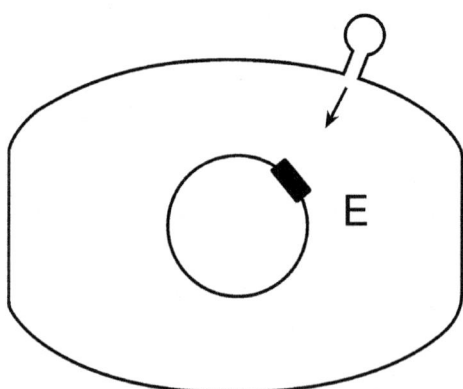

F. Maintained as such for several generations. occasionally phage DNA gets detached from the bacterial chromosome

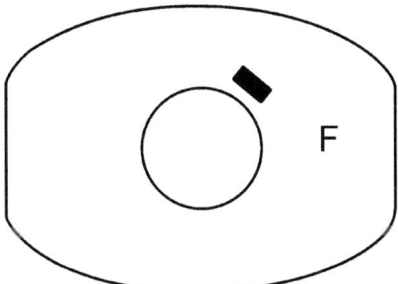

G. The bacterial DNA and takes control of the bacterial cell behaving like a virulent phage causing lysis of the cell with the release of mature bacteriophages

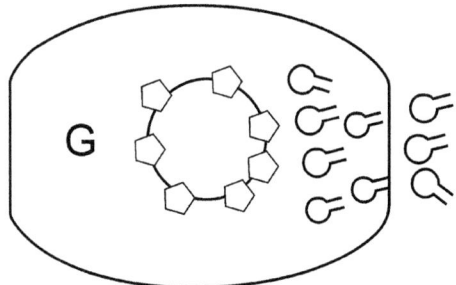

At is a soil bacterium which infects the crown region of dicotyledonous plants (monocots are resistant) through mechanical wounds to produce crown galls. A.t harbours large extrachromosomal megaplasmids. Most of the genes required for tumour formation are located on one such 180 kb megaplasmid designated as Ti plasmid, the letters Ti denoting the tumour inducing ability of the plasmid. The Ti plasmid contains a tumour inducing region (T-DNA) which also carries genes for the synthesis of two growth hormones, the IAA and cytokinins and the genes controlling the synthesis of a group of amino acid derivatives known as opines (nopaline and octopine).

The expression of bacterial genes controlling growth hormone production is not controlled by the plant but is necessary for the development of crown gall symptoms. The opines synthesized serve as nitrogen and carbon sources for the bacterium. However, the genes controlling the growth hormone production and opine synthesis are not essential for transferring or integration of T-DNA into the host plant cell genome.

The Ti plasmid has also a cluster of about 8 genes known as virulence genes *(vir* genes). This gene cluster of 35 Kb DNA is necessary for the recognition of susceptible cells on the plant surface, to excise T-DNA from the plasmid and transfer the T-DNA region to the host cell. The *Vir* genes are activated only by contact with cell metabolites released by the wounded plant and they are not functional or expressed in pure A.t cultures grown on synthetic media.

There are two border regions to T-DNA (LB and RB) which are known to contain genes involved in the secretion of the enzyme endonuclease that scissors off the T-strand from the Ti plasmid. Thus the genes encoded within the T-strand have all the appropriate signals for efficient transcription and translation in their eukaryotic host.

The process of infection begins with the bacterial surface components of A. t. recognizing the plant surface which is susceptible to the attack by the pathogen, followed by a process analogous to bacterial conjugation whereby a single strand of T-DNA (the T-strand) is transferred to the plant cell probably through pores in the cell wall.

Within the host cell, several copies of T-DNA are inserted at single or multiple sites in the host chromosomes and function as typical eukaryotic chromatin.

The earlier procedure to identify transformed cells was to develop plant galls or tumours from such transformed cells. Alternatively, the transformed cells were grown in hormone independent cultures. However, in later experiments, modified T-DNA and Ti plasmids have been used to facilitate rapid experimentation and development of transformed plants. In the modified T-DNA, the genes coding for phytohormones and opines have been excised because, as stated earlier, these genes are not concerned with the transfer and integration of T-strand into the host genome but are essential for the manifestation of crown-gall symptoms.

In essence, deleting these genes from T-DNA results in the elimination (disarming) of the oncogenic (tumourous) phenotype. Foreign DNA can be inserted within the right and left border regions of an appropriate plasmid, thereby enabling the delivery of large multigenic segments of DNA into plants. In this way T-DNA region has been engineered (disarmed) to eliminate the tumour inducing genes. Selection of transformed cells has been facilitated by engineering a variety of antibiotic markers selectable in plants such as resistance to kanamycin and gentamycin.

Ti plasmids have also been engineered in other Agrobacterium spp. and E. coli to facilitate efficient plasmid replication and selection. Such manipulated plasmid vectors could be used directly in transformation steps obviating the need to repeatedly infect plants with A. tumefaciens cells.

Procedures for Agrobacterium Mediated Transformations

Several procedures have been followed till to date: (a) through conventional gall formation on wounded plant stems or.leaves that are not amenable to monitoring with disarmed vectors, (b) Co-cultivation of *Agrobacterium* with plant protoplasts, a method that has been successful in plants where regeneration of plants from protoplasts has proved to be successful and (c) tissue transformation with ex-plants (leaf, cotyledon section or somatic embryos) inoculated with *Agrobacterium,* a procedure that has proved faster than protoplast regeneration.

When shoots are formed in regenerated plants, the plantlets are transferred to potted soil for acclimatization to natural surroundings.

The introduction of *nif* genes controlling nitrogen fixation in some microorganisms, into higher plants other than legumes is an experimental strategy envisaged by several research groups.

Direct DNA-Transfer Technologies

Most monocotyledons like cereals and sugarcane are not amenable to *Agrobacterium* technology because they are resistant to infection by the bacterium. Alternative technologies are being developed which include the introduction of foreign DNA into plant protoplasts mediated by polyethylene glycol and poly-L-ornithine, electroporation (use of electric current), calcium phosphate coprecipitation, liposome fusion, microinjection and particle bombardment. Regeneration of plants from transformed protoplasts is labour intensive and subject to the possibility of develop-ment of somoclonal variations.

The use of 'microprojectiles' or 'particle bombardment' is a procedure for delivering foreign DNA into plant cells. The technique involves coating small gold or tungsten beads with plasmid DNA and propelling the beads to intact cells, embryos or differentiated tissues using high velocity particle , guns' or electrical discharges. This technique has been useful in delivering DNA into the nucleus and mitochondria of yeast, nucleus and chloroplasts of *chlamydomonas,* epidermal tissues of *Allium cepa* and suspension cultures of maize.

A major limitation of this technology arises from the paucity in the number of particle guns available and the high cost in building them.

Antisense RNA Strategy

The basic idea in antisense strategy is to block the expression of a particular gene product with the help of a transgenic construct containing the gene or part of the gene with the transcript in reverse orientation with respect to the promoter. This procedure results in the formation of a complementary RNA rather than the normal sense RNA.

This antisense RNA binds with its homologous sense RNA, preventing translation and/ or facilitating degradation or accumulation of a gene product by 90 to 99 per cent, thus creating a phenotypic mutant. There have been many reports of insertion of antisense construct of polygalacturonase enzyme into tomato plants. The resulting transgenic tomato plants had low levels of this enzyme and hence their fruits ripened slowly thereby increasing shelf life. Antisense RNA technology has also been used with success towards minimizing viral diseases.

Frost Control Biotechnology

Frost injury to plants is caused at temperatures less than 0°c. Two kinds of frost injuries have been recognized: those occurring above minus 5°C and those occurring when temperature drops below-5°C. Plants resist frost by restricting freezing to intercellular spaces and adjusting the water potential intracellularly to reach equilibrium with the ice formed in inter-cellular spaces. Frost injury takes place when this equilibrium is upset and the rate of intracellular ice format on exceeds that of intercellular spaces followed by death due to disruption of cell membrane properties.

Frost tolerant plants, however, have endogenous ice-tolerance mechanisms.

How does frost sensitive plants overcome frost injury. There appears to be a super cooling mechanism in such plants to avoid ice formation. Ice nucleation or initiation of ice embryos is caused by the orientation of water molecules by organic and inorganic substances. Several plant-associated bacteria are highly active in ice nucleation and may be responsible for frost injury and the biotechnological implications of this activity have been studied with reference to frost injury at temperatures above minus 5°c.

Several strains of bacteria such as *Xanthomonas campestris,* *Pseudomonas viridiflava,* P. *fluorescens* and *Erwinia herbicola* inhabit the epidermal crevices and hairs of leaf surfaces and are active in ice nucleation at temperatures above minus 5°C.

It should be noted that the efficiency of ice nucleation activity of these bacteria differs not only with the strains of bacteria but also with the plants harbouring them. The population density of ice nucleating bacteria is variable on plant surfaces depending upon the species of plants and the environmental conditions under which they grow.

These epiphytic bacteria are known to occur on frost resistant as well as frost susceptible plants and are known to be killed by disinfectants and V.V. light.

The genes conferring ice nucleation characteristic have been partially characterized from P. *syringae,* P. *fluorescens* and E. *herbicola* and are known to be a single contiguous region of approximately 4000 bp. They have been cloned in *Escherichia coli.*

Streptomycin and oxytetracycline as well as *copper* hydroxide applications to leaf surfaces reduced the incidence of frost injury, despite the fact that dead cells were also known to be active as ice nucleating agents. Likewise, non-ice nucleation-active bacteria can also reduce the population of ice-nucleating ones both in the green house and the field by limiting nutrients to the ice nucleating bacteria.

For example, non-ice nucleation-active (Ice mutants of P. *syringae* reduced the population size of ice nucleation active parental strains of P. *Sryringae* that were co-inoculated on pretreated plants.

Field trials have been conducted to understand the competition behaviour of ice" strains of P. *syringae* by inoculating these strains to potato plants. Ice' strains dominated the leaf surfaces for the first 4-6 weeks after inoculation. The population of Ice+ strains on plants colonized by Ice- P.*syringae* strains was significantly decreased in comparison with uninoculated plants.

The incidence of frost injury to potato plants inoculated with Ice strains was significantly lower than uninoculated control plants in natural field frosts in a field experiment in California.

The use of microorganisms for competitive control of frost injury have been found to be effective only when applied to young vegetative plants in the field because such young vegetation may not have been extensively colonized by other epiphytic microflora.

Minimal occurrence of extraneous nicroflora on leaf surfaces which can be achieved by the application of bactericides such as cupric hydroxide can result in micro-habitats most conducive for the functioning of Ice- P. *syringae* in large numbers on the leaf surface so that the Ice-strains can effectively compete for nutrition with frost inducing naturally occurring wild strains of P. *syringae.* In-tegrated chemical and biological control measures to contain frost injury to plants appear to be a desirable approach. In addition, co-application of copper resistant Ice strains of P. *syringae* and cupric hydroxide has also been considered as an attractive proposition to control frost injury.

Virus Resistance in Transgenic Plants

There have been many reports about the development of transgenic plants which have become resistant to virus infections. These plants showed resistance to virus infections when they were transformed with sequences related to several gene functions.

Some of the areas where successes have been seen related to sequences concerning viral capsid protein, viral move-ment protein, antisense RNA, antibody-mediated resistance, interferon-related genes and host genes involved in plant protection.

Viruses are biochemical complexes consisting of a RNA or a DNA genome packaged into a protein capsid which mayor may not be sur-rounded by a membrane envelope. The protein coat 'covered genome is referred to as the nucleocapsid. The proteins on the surface of the capsid and envelope determine the interaction of the virus with the host and elicit the protective immune response against the virus. Some virus particles also contain enzymes required to facilitate the replication of the virus. The tobacco mosaic virus (TMV) is an example of a virus with helical symmetry whose capsomeres (many protein subunits of the capsid) appear as projections that are assembled on the RNA genome into rods extending to the length of the genome. In other viruses the capsomere arrangement is cubical or icosahedral enclosing its nucleic acid component. Engineering resistance in plants involves either countering the capsid properties or disrupting the virus replicating mechanisms in the hot.

Transformation of Sequences Related to Viral Capsid Protein

Coat mediated resistance is the expression of a gene that causes the transformed cell and regenerated transgenic plants to produce the coat protein (CP) of a virus thereby conferring resistance to infection in the transgenic plant. Plants that accumulate large amounts of coat protein escape/mini- Elementary features of typical virus particles: A-Enveloped icosahedral virus; B-Naked icosahedral virus; C-Enveloped helical virus and D-Naked helical virus maize virus infection.

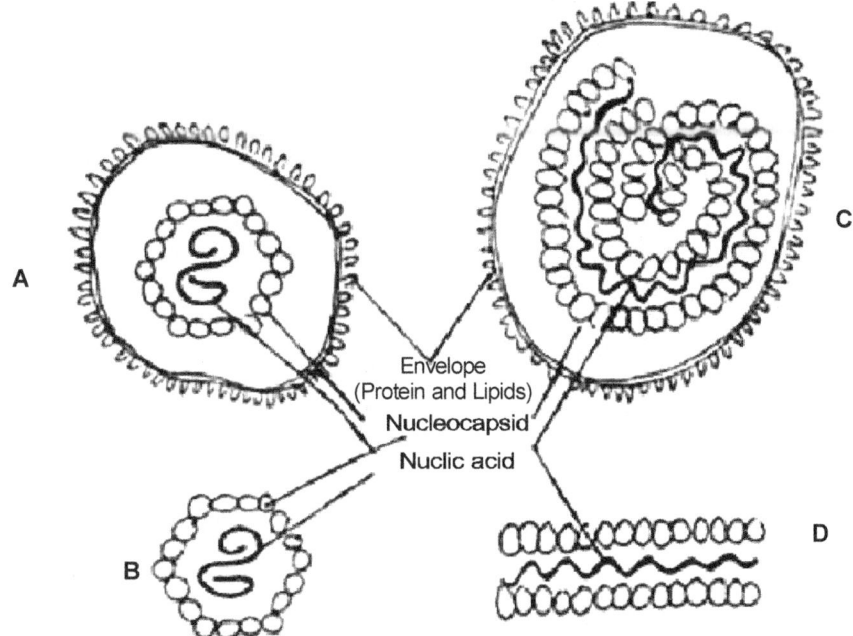

Figure 2 : Elementary features of typical virus particles.

The CP mediated resistance was first reported in tobacco mosaic virus (TMV) in 1986. Subsequently several instances of resistances for several virus diseases of plants have been reported which include tomato mosaic virus, alfalfa mosaic virus, cucumber mosaic virus, tobacco streak virus, potato virus x and y, tobacco etch virus, tobacco rattle virus, potato leafroll virus, potato virus S.

A point which needs to be stressed is that CP mediated resistance does not provide immunity to transgenic plants; for instance TMV resistant plants that expressed CP gene were susceptible to inoculum levels of 10mg/ml while plants that were not transgenic got infected even at inoculum levels of 0.001-0.01 mg/ml.

Despite the fact that the mechanism involved in CP mediated resistance has not been clearly understood, some kind of inferences can be made from studies on TMV diseases in tomato and tobacco. Firstly, the transgenic plants are protected due to reduction in the number of infection sites that means fewer number of plants may get infected than controls. Secondly, the transgenic plants are less likely to develop systemic infection and thirdly, such plants may produce less virus particles than the ones not having the resistant gene.

Transformation of Sequences Related to Viral Movement Protein

A gene product of TMV was the first to be directly identified and investigated in viral movement. Plasmodesmata are the portals between adjacent host plant cells responsible for viral movement.

A gene product (viral protein) has been found to accumulate at or near plasmodesmata of TMV infected transgenic plants especially in the cell wall fraction, which directly or indirectly changes the plasmodesmatal permeability to allow the passage of TMV particles measuring 18 x 300 nm (30-KDa) while in the normal tobacco plants only low molecular weight compounds less than 1.5-2.0 nm could pass through.

However, microinjection of recombinant viral protein into plant cells resulted in the enhancement of plasmodes-matal expansion upto 9nm limit, that was insufficient for viral movement to take place between host cells and hence an alternate explanation had to be found.

It is known that the viral protein also binds to RNA and hence it has been conceived that one part of the viral protein binds to TMV virus RNA and shapes into a complex measuring about 2nm and the other part of the protein binds to plasmodesmata to increase their permeability character and thus enable the virus to pass through plasmodes-mata.

Resistance Conferred by Antisense RNA

Antisense regulation of gene expression is a natural phenomenon concerning the specific expression of a nucleotide strand which is negative to a certain gene transcript capable of intervening the expression of that gene at different levels. Antisense RNA may bind to the loop initiated by RNA polymerase in the nucleus and interfere in the initiation of transcription.

It may function in the cytoplasm by hybridizing with mRNA leading to translation arrest by preventing the binding of ribosomes to mRNA. Antisense regulation of gene expression has been successfully exploited in plants. Some examples are inhibition of flower pigmentation by antisesnse for chalcone synthase, intervention in the expression of ribulose biphosphate carboxylase in tobacco and polygalacturonase in tomato.

Most plant viruses replicate in the cytoplasm and do not go through any nuclear phase and therefore antisense RNA may either intervene in translation or promote mRNA degradation. Antisense mediated resistance has been achieved in transgenic plants against several plant viruses including TMV. This mode of resistance appears to be milder than coat—-mediated protection.

Antibody-Mediated Resistance

Antibody-mediated protection (akin to hybridoma-derived monoclonal antibodies) from only two plant viruses (TMV and artichoke mottle crinckle virus) are known. The mode of action of antibodies appears to be through neutralization of viral surface proteins in a way that intervenes the initial establishment of infection.

3

Role of Bacteria and Actinomycetes in Biological Control of Pests and Diseases

MICROORGANISMS We find that more life is living below the surface than above in most environments. Soil are put into six groups: earthworms, nematodes, arthropods, bacteria, fungi, and protozoa with each having their own function. They do play an important role in plant health and water.

The cycle of nutrients going through our environment are mainly driven by these microorganisms. They maintain:

1. Decomposition
2. Mineralization
3. Nitrogen cycling
4. Storage and release of nutrients
5. Carbon cycling
6. Take the pollutants out of the water before it reaches underground or surface water.

The Location, Living Conditions, and Functions of Microorganisms

Actinomycetes Effective microorganism, they are responsible for distinctive scent of freshly exposed, moist soil. Prefer neutral to alkaline soils; high oxygen requirement; prevalent in dry regions. They release carbon, nitrogen, and ammonia during decomposition of organic matter. They help form humus. Associate with non-leguminous plants like bitterbrush to fix nitrogen and make it available to other plants in the area. Some nitrogen may be unusable without the bacteria converting it to a form that can be used.

Bacteria Microorganisms found all over the world and even found down in the earth as far as 1 mile. It thrives under most conditions but found near plant roots, an important food source.

Feeds on organic matter; encourage organic and inorganic chemical reactions that affect plant growth; and fix nitrogen from air in soil.

Soil Bacteria

What bacteria lack in size, they make up in numbers. They are tiny, one-celled organisms. A teaspoon of productive soil generally contains between 100 million and 1 billion bacteria.

Has Four Functional Groups

1. Most are decomposers that consume simple carbon compounds. By this process, it convert energy in soil organic matter into forms useful to the rest of the organisms. A number of decomposers can break down pesticides and pollutants in soil. Decomposers are especially important in stopping or retaining, nutrients in their cells, thus preventing the loss of nutrients, such as nitrogen, from the rooting zone.
2. A second group are the mutualists that form partnerships with plants. The most well-known of these are the nitrogen-fixing bacteria.
3. The third group is the pathogens.
4. A fourth group, called lithotrophs (literally meaning rock eaters) or chemoautotrophs (which are able to synthesize all of the organic compounds they need from inorganic raw materials in the absence of sunlight.), obtains its energy from compounds of nitrogen, sulfur, iron or hydrogen instead of from carbon compounds.

Functions

- They perform important services related to water dynamics.
- nutrient cycling
- disease suppression.
- many organisms will compete with disease-causing organisms in roots and on above ground surfaces of plants.

Important Bacteria

Nitrogen-fixing The plant supplies simple carbon compounds to the bacteria, and it converts nitrogen (N2) from air into a form the plant host can use. When leaves or roots from the host plant decompose, soil nitrogen increases in the surrounding area.

Nitrifying change ammonium to nitrite then to nitrate – a preferred form of nitrogen for grasses and most row crops. Nitrate is leached more easily from the soil, so some farmers use nitrification inhibitors to reduce the activity of one type of nitrifying (scavenging potentially toxic nitrogen compounds from their surroundings, including: ammonia and nitrite). Nitrifying it are suppressed in forest soils, so that most of the nitrogen remains as ammonium.

Denitrifying convert nitrate to nitrogen (N_2) or nitrous oxide (N_2O) gas. Denitrifiers are anaerobic, meaning they are active where oxygen is absent, such as in saturated soils or inside soil aggregates.

Actinomycetes are a large group of bacteria that grow as hyphae like fungi. They are responsible for the characteristically "earthy" smell of freshly turned, healthy soil. Actinomycetes decompose a wide array of hard-to-decompose compounds and are active at high pH levels. Fungi are more important in degrading these compounds at low pH. A number of antibiotics are produced by actinomycetes such as Streptomyces.

They are found where?

· It is more competitive when easy-to-metabolize substrates are present such as fresh, young plant residue and the compounds found near living roots.

· They are especially concentrated in the region next to and in the root.

· Some believe that plants produce certain types of root exudates to encourage the growth of protective bacteria.

One Celled Organisms that Encourage Plant Growth

Certain strains of the soil bacteria have anti-fungal activity that inhibits some plant pathogens (A disease-causing organism). Those strains can increase plant growth in several ways. They may produce a compound that inhibits the growth of pathogens or reduces invasion of the plant by a pathogen. They may also produce growth factors that directly increase plant growth. The plant growth-enhancing organism occur naturally in soils, but may not be in high enough numbers to have a great effect. Soon, gardners may be able to coat seeds with this anti-fungal to ensure that the bacteria reduce pathogens around the seed and root of the crop.

Earthworms are the reason for a huge industry today. People are raising & selling earthworms all over the world, especially wholesale earthworms. The work of the earthworms in the soil is what earthworms eat. Producing castings that are equivalent to compost as an organic fertilizer and soil supplement.

Fungi Basically 2 types of fungi-mycorrhizal and normal. Fungi thrive in well-drained, neutral to acidic, aerated soils. Normal fungi help decompose the organic matter in litter and soil but play less of an overall role. Mycorrhizal fungi help develop healthy root systems by growing on plant roots. The fungus is actually a network of filaments that grow in and around the plant root cells, forming a mass that extends considerably beyond the plant's root system. This essentially extends the plant's reach to water and nutrients, allowing it to utilize more of the soil's resources.

Mode of Entry of Microbial Pathogens on Insect Pests and their Mechanism of Control

In Coffee

In the early 1870's the coffee leaf rust caused by the fungus Hemelia vastatrix, wiped out coffee plantations in Srilanka. In spite of the pioneering work of the English pathologist, H.Marshal Ward, no biological or chemical cure could be found to save the coffee plantations. Hence, the coffee production was abandoned in Srilanka and shifted to the Western hemisphere. In the early days, the British pioneers who opened up coffee plantations restricted themselves in growing the Arabica variety. Coffee cultivation reached its peak in the 1860's and it was at this point that the major outbreak of pests and diseases took place. Hemileia vastatrix—coffee leaf rust, root rot, berry blotch, collar rot, white stem borer –Xylotrechus quadripes, mealy bugs, green bugs, berry borer, shot hole borer, red borer, thrips etc. In search of remedies, number of toxic chemicals was tried out and even today the list of chemicals is ever growing.

Chlorinated hydrocarbons, mercury laced compounds, lead, Some of these chemicals were developed in the Western world and directly applied under Indian conditions without screening; resulting in resistance build up. Over the years the target organism has evolved new races or mutations which are tolerant to more powerful chemicals with inbuilt resistance. Planters never bothered to use pesticides diligently. We are suggesting a simple, yet effective approach of integrated pest management (IPM) in controlling and not eliminating pest and disease incidence.

Microbial Inoculants

Microorganisms can be effectively used as a safe and alternative method in controlling insect pests. Microorganisms that affect insects are termed ENTOMOPATHOGENS. Microbial inoculants or bio agents are Entomogenous microorganisms and or their products that cause pathologies and are usually fatal to their insect hosts. Almost 1500 insect pathogens have been isolated world wide but unfortunately less than one per cent has been exploited for commercial value. About 5 or 6 of them have been tried commercially. Their role includes disease initiation in target insects or to suppress harmful insect population directly or in combination with chemical insecticides or pesticides. Entomopathogens multiply in nature if the environment is conducive and cause infections on the target organism. Scientists have proved that microbial inoculants pose little human or environmental damage.

Prerequisities of Microbial Inoculants :

- Bio inoculants should be highly virulent to the target organism.
- Should be stable to the stress of nature.
- Virulence should be confined to the species of insects for which it is treated (Should be host specific).
- Should neither damage the crop, nor harm beneficial insects and micro flora and predators and animals.
- Should be easily mass culturable.
- Should be stable for a long period of time.

Advantages of Microbial Inoculants :

- They are environment friendly and leave behind no toxic residues.
- Most of them target the specific insect and in turn protect beneficial insects.
- Most inoculants are easily culturable in the lab, with minimum space.
- Inexpensive to produce large quantities of inoculum.
- Slowness in developing resistance to microbial pathogens.

- Can control insect in cavities where chemical insecticides cannot reach.
- In a way, we try to mimic nature by releasing them into an open environment.

Microbial inoculants can be used in two ways:

A. SHORT TERM CONTROL: For a particular season (annual) or by using highly virulent pathogens.

B. LONG TERM CONTROL: Generally used for perennial crops and less virulent pathogens are preferred.

Disadvantages of Microbial Inoculants :

- Necessity for careful and correct time of application.
- Host specificity of most pathogens, narrows down its use.
- Necessity of maintaining a pathogen in a viable condition, until the insect is contacted.
- Difficulty in producing some obligate and facultative pathogens on a large scale.
- Requirement of favourable environmental conditions for the pathogen to act, multiply and execute its mode of action.
- Tendency of dead insects remaining attached to the host.

Bacteria

There are approximately 100 species of bacteria pathogenic to insect pests but hardly one per cent among these is used as bio control agents. Bacteria belonging to the family Bacillaceae, Lactobacillaceae, Micrococacaceae, Pseudomonadaceae and Enterobacteriaceae are generally found pathogenic to insect hosts. For E.g. the most commonly used Bacteria in biological control of insect pests, Bacillus thuringiensis which is a spore former, was first isolated by a German by name Berliner in 1915 and was subsequently known to control several insect pests.

Even today it is known as one of the most virulent bacterial pathogen which kills many larval forms of lepidopterous insects. It affects more than 175 larval species that include some of the most important economic pests. It can also be mixed with a number of commercial insecticidal formulations and is gaining ground in integrated pest management. There are many other species of bacteria which are used in biological control of insect pests such as Aerobacter aerogenes and Escherichia coli, Bacillus popillae, Bacillus sphaericus,

Bacillus moritai, Bacillus lentimorbus are already used in commercial preparations. These bacteria form a toxic crystal protein which acts as a stomach poison inside the insect host resulting in the quick death of the insect. The above mentioned genus of bacteria attack mostly members of the insect family belonging to Hemiptera, Diptera, Coleoptera and Lepidoptera. Some species of bacteria in addition to forming toxic spores produce crystalline inclusions within the sporulating cell which aids in killing the target insect. Bacillus popillae and Bacillus lentimorbus when sprayed in the field invade the insect and gains entry into the blood system. This acts as a good medium for the growth and proliferation of the microorganism. The bacteria sporulates causing imminent death of the insect and once death occurs the bacterial spores are released into the atmosphere and soil and remain there for extended periods of time. These spores may in turn infect newly hatched beetle larvae and in turn infect it.

Fungi

Fungi belonging to the family Deuteromycetes and Entomopthora causes mycosis in insects. More than 800 fungal species comprising of 125 genera have been reported to infect insects. Probing a little deeper into this vast number of fungal species reveals that most harmful pests are associated with one fungi or the other. However, more work needs to be done in developing fungal biocontrol agents. They attack mostly members of the family Hemiptera, Lepidoptera and coleoptera. Approximately 100 genera of fungi are pathogenic to insects. Fungi penetrate the body directly. They invade insect's body walls and spread in the softer tissues of insect body. Fungi infected insects show a thick mycelial mat covering the entire body of the insect. In nature the insect may either be exposed to the thick mycelial mat found on plants, leaf or soil or may also be exposed to toxic fungal metabolites either by cuticular contact by ingestion or by injection into the blood stream known as haemocoel.Beauvaria bassiana the white muscardine fungus is a potential bio inoculant against 700 species of insect pests. The green muscardine fungus Metarhizium anisopliae has been reported to effectively control a large number of target organisms.

Viruses

Insect viruses fall into five major groups. They belong to the family Baculoviridae.

1. Nucleopolyhedrosis Viruses {NPV}

2. Granulosis Viruses {GV}

3. Cytoplasmic Polyhedrosis Viruses {CPV}

4. Non Inclusion Viruses {NIV}

5. Entomopox Viruses {EPV}.

More than 800 viruses have been isolated that target ARTHROPOD insect pests. Since insect viruses are absent in vertebrates and plants, they are safe not only to human beings, but also to no target organisms. Insect viruses are obligate parasites and hence need to be cultured in living hosts. The target pest is reared in the laboratory or factory and mass multiplying the virus by inoculating it. Nucleopolyhedrosis virus is the most commonly used for insect pest control. When viral biocide is sprayed in the field, the viral particles get attached to the foliage and the insect larvae feed and in turn ingest the polyhedral inclusion bodies (PIB). The pib in turn multiply inside the midgut and bring about larval mortality. These protein crystals are insoluble in water and are produced in geometrical proportions and after killing the insect, get released into the surrounding atmosphere. The crystals can reinfect live insects and retain their infectivity even after long storage outside the living tissue of plants.

Protozoa

Protozoa belonging to the family Nosematidae has shown great promise as a bio inoculant. Approximately 375 species of protozoa have been found to be pathogenic on various harmful insects. Nosema locustae has been tried out in the control of grasshoppers.

Nematodes

Nematodes parasitic on insects like rootworms, grubs, caterpillars have been recorded from early times.

Mode of Action

The two main factors in controlling the disease are by INVASIVENESS AND INITIATION of disease. The site of entry of bio inoculants may be broken or bruised integuments or by feeding habits of the insects or other openings. When attacking any insect the quantity and quality of inoculum is important. In essence the control of the disease depends on the capacity of the microbial inoculum to rapidly establish itself and multiply in large numbers, resulting in severe infection of the pathogen concerned. The time lag between the entrance of microorganisms into the body of the insect and the incubation period within varies from pathogen to pathogen and insect to insect. Proliferation and disease control are controlled by environmental factors to a large extent and internal host characteristics to a small extent. The mode of action may be due to the production of ENDO or EXO toxins.

Broadly bioagents which attack insect pests can be classified as follows.

1. *Obligate Bioagents:* These bio agents in nature are associated with a specific insect disease and are difficult to culture in artificial media. E.g. Bacillus popilliae causing milky diseases of white grubs and Bacillus larvae causing American foulbrood of bees.

2. *Spore Forming Bio Agents:* The protein crystals formed by these sporulating bio inoculants are highly toxic to the target agent. E.g. Bacillus cereus produces such toxins.

3. *Determinative Bioagents:* The bio agents get activated; once inside the host. They multiply in the haemocoele of insect and produce lethal septicemia.

4. *Facultative Bioagents:* The host tissue is damaged by the invasion of the bio agent but they are not obligate parasites.

5. *Predaceous Bioagents :* The bio agent either secretes a sticky substance or builds up.

We are certain that mankind has not provided adequate resources in terms of money, research and time in search for naturally occurring or safe alternate methods of pest control. It is simply because our mind is tuned towards shortcuts and profits. The use of bio inoculants or their by products is under exploited. The dangers of chemical poisons are there for all to see, right from water bodies to food chains, all of nature too is affected. Due to the accumulation of toxic metabolites in the air as well as soil system, it will result in creating immunity to dangerous pests. The genetic population of future generation will be the one, where less susceptible individuals are left to breed.

They leave behind untold miseries for the future children of this planet including wild life habitat, birds, fish, mammals etc. Their long term effect on the environment cannot be quantified by scientists sitting in the laboratories. We need a more intelligent way of using chemicals. On the other hand if we can only apply ourselves to the fact that evolution has all the answers to our problems, it will only be a short time before we discover natural agents which will aid in biological control of pests and diseases.

This approach will make the world a better place for all to live and share. Microbial Inoculants result in much greater protection for both flora and fauna. It is like a NATURAL system taking place in nature. We do not imply that modern technology should be abandoned. On the contrary we need to focus on Integrated Pest Management which will give us a better hope for tomorrow. This technology has been successfully tried out in many developed countries like the U.S.A., Canada, and Germany where it is used on several agricultural crops, trees and ornamental shrubs. It has been found to be economically viable and most importantly environment friendly. Microbial inoculants open up a world of opportunities. They create living spaces which bring about harmony in nature. We are blessed people to live on this bio diverse rich planet. Chemical control of stored product insect pests has been the most efficient and effective means of protection of stored produce. However, with the increasing cost of inorganic chemicals and their known hazards to the environment an integrated means of control has been widely adopted. The search for botanical insecticides could supplement the expensive petroleum based chemicals. Botanical preparations have long been used for protection of stored produce by small scale farmers in oriental countries such as India where the neem tree has been used extensively.

The success of these small farmers serve as the impetus for exploring the utilization of indigenous resources for small scale product protection and for possible industrial scale product protection. Several species of plants are now being screened for source of active ingredients against stored product insect pests commonly used are made of polyethylene or polyvinyl chloride film having a thickness of 0.1 mm (.004 inch) weighing 100 grams per square meter. After the completion of the fumigation it is necessary to collect the tablet residues or expended sachets. These materials can be disposed properly by swamping it in a soapy water until bubbling ceases (in open air) and then buried at least 30 cm below the soil surface. Under no circumstances should large quantities of expended residues be placed in a heap prior to being buried, because of the possible danger of ignition.

Formulation Available and Dosage Rate

Phosphine (PH3) is commonly available in the form of tablets, pellets and sachets containing aluminum phosphide.

- Each tablet weighs 3 g and after complete decomposition 1 g of phosphine is released.
- Each pellets weighs 0.6 g i.e., 1/5 of tablet weight) and on complete decomposition, 0.2 g of phosphine is released.
- Each sachet weights 34 g and on complete decomposition 11 g of phosphine is released.

The following equivalents, based on commercial formulations containing aluminum phosphide as the phosphine source are given to simplify dosage calculations.

1 g phosphine = 1 tablet = 5 pellets = 1/11 sachet

Under Philippine conditions, the following rates of phosphine application are recommended.

15-45 tablets* per 100 cu. Ft	2-5 tablets per metric ton	15-20 tablets per 1000 cu ft
Or	or	or
3-5 sachets ** per 1000 cu ft	1 sachet per 1-3 metric ton	2-3 sachet per 1000 cu. ft.

* Recommended exposure time 3-5 day when tablet is used

** Recommended exposure time 3-6 days when sachet is used.

Identification of Active Principles

Neem Leaves and Fruits

The constituents of the neem leaf powder were obtained by Tirimanna (1985) using two-dimensional thin layer chromatography. As a result of a strong saponification procedure, it is only the very stable compounds, which were observed on the chromatoplate. The compounds violaxanthin, butoin, 3carotene, zeaxanthin, ant heroxanth in, cryptoxanth in and neoxanthen were identified.

Seven epoxy compounds have also been isolated from fruits of Melia azedarach (Kraws et al., 1981). These compounds were rather stable and maybe of significance to scientists in search of a stable compound having insecticidal properties. This is relevant in view of the fact the neem leaf loses insect repellent property with age.

One sterol and 3 stable ketones were found, together with six very stable phenolic compounds. The occurrence of phenolic compounds in the neem plant is already well documented. Phenolic constituents are also recorded as antihelminthic factors (Taniguchi, 1960). The identification of quercetin in the neem leaf accounts for its antibacterial and antifungal properties and hence the curative properties of leaves in cases of sores and scabies.

T. diversifolia (Wild Sunflower)

The active fraction D from this plant contains alpha-lactone with a hydroxyl group attached either to the lactone ring or to the alkyl substituent. The unsaturation may be present in the ring itself or alpha to the carbon as indicated by the strong IR absorption at 1670 cm' and by the failure of the sample to decolorize bromine in carbon tetrachloride.

Insecticidal Activity

Black Pepper (Piper Nigrum) and Red Pepper (Capsium Frutescens)

They found the crude and semi-purified extracts of black pepper topically applied on the insects toxic. The admixture treatment of grains showed the ground black pepper (GBP) more toxic than the semi-purified and crude extract as residual contact and stomach poisons against the corn weevil. It was also found resisually toxic for 2 months on the saw-toothed grain beetle, lesser grain borer and corn weevil.

The ground red pepper (GRP) and GBP were residually toxic to the bean weevils for 2 and 6 months, respectively. The effectiveness of red pepper on bag rice under warehouse condition (natural infestation) at average temperature of 25.70-27.90°C was further evaluated. The major stored product insect pests were corn weevil and red flour beetle.

Three methods of applications were done: mixing powdered and whole fruit of RP with rice grain or in sachet with doses of 1200 ppm or 1800 ppm and spraying the bag with either 5, 10 or 20% of the extract. All methods of application and dose levels effectively protected the bagged rice from infestation for no longer than two months storage.

Neem (Azadirachta indica A. Juss)

The repellent and antifeedant effect of seed powder and oil of neem on five species of stored product pests was investigated by Akou-Edi (1985). Laboratory trials in Togo using red corn treated with neem oil at concentrations of 1, 2, 4 & 8 ml/kg or with seed powder at 20, 40, 80 g/kg infested with confused flour beetles and corn weevils showed significant difference between the treated and untreated samples. As a rule, the effect of neem oil and need powder increased with higher concentrations but difference between concentrations was not significant on both test insects.

Atudy on the effect of neem seed kernel and leaf powder on the development of Callosobruchus maculatus (F.) on 3 bean seed species treated at the rate of 0.5, 0.1 and 2.0 parts per 100 parts of seeds showed reduced oviposition capacity, lengthened larval and pupal period, reduced percent adult emergence, longevity and growth index. Among mungbean, bush bean and cowpea bean, the latter was found to be the most preferred food for the development because it supported normal development and enhanced population growth.

Studies on the effect on neem kernel extract (NSKE) on metamorphosis of Ephestia kuehniella (mill moth) and which established a concentration effect curve were undertaken by Maurer (1985). A methanolic extract of sundried neem seeds was applied at a concentration of 4 ppm on 4-5th instar larva. The stage of development that a larva had reached was determined by measuring the size of the head capsule. The average size of the head capsule at each larval stage was determined in a preliminary experiment on untreated larva under the same conditions (30°C, 82-86% RH). A 4 ppm concentration and less resulted in the insertion of an additional larval instar between the fourth and fifth larval stages. Hence, metamorphosis was prolonged after treatment with neem seed kernel extract. The mortality induced by concentrations of 4 ppm NSKE or higher, occurred first during the molting phase between different larval stages. No feeding deterency was observed.

Lageundi (Vitex Negundo)

The protectant effect of whole and powdered leaves of lagundi on stored corn against corn weevil was evaluated by Bhuijah (1988) for a period of 6 months. Two level of dosages were used (1 & 5%) for both whole and powdered leaves on 5 kg of corn. After 30 days of storage, the protectant effect of 5% lagundi leaf powder and 1% whole leaf was lost as shown by increased infestation of all lagundi treated samples. Only neem seed oil treated corn seeds remained protected and had low percent infestation throughout the study. Apparently the insecticidal constituent of the lagundi leaves were not stable, hence the repellent property were readily lost.

Vegetable Oils Controlling Storage Insect Pests The practice of adding a little vegetable oil to stored rice or legumes for protection against stored-insect pests is well known and well established in oriental countries like China, India and Indonesia. Recently the practice of protecting stored produce with oil has spread and has been adopted in Africa and South America. Recently Van Rheenen (1983) pointed out the applicability of this method of protecting storage grains to supplement safe chemical formulations.

The mode of action, appropriate dosages and duration of efficacy of oils have been investigated by various workers on storage insect pests. Differences between crude and purified oil have been studied and crude oil has been found to be a better protectant while the triglyceride oleic acid combination was found the most effective (Schoonhoven and Hill (1981).)

The amount of oil needed for the control of most storage pests vary from 2 cc/kg seed to 15 cc/kg seed (Cruz and Cardona, 1981) depending on the level of infestation.

Table 1 : Percent infestation of treated corn grains at different storage periods at room temperature

Treatments	DAYS					
	30	60	90	120	150	180
Control	2.04a	10.60a	22.91a	61.97a	80.27a	88.51a
1% lagundi leaf powder	1.67a	11.24a	18.26a	58.51a	78.11a	90.22a
5% lagundi leaf powder	0.90b	6.79a	14.42a	65.73a	77.51a	90.15a
1% lagundi whole gried leaf	1.52ab	5.48ab	15.69a	62.20a	74.79a	38.58a
5% lagundi whole dried leaf	2.55a	10.80a	19.05a	57.12a	70.73a	88.81a
Actellic	0.38bc	0.33b	0.17b	5.45b	5.24b	14.19b
Neem seed oil	0.25c	0.06b	0.00b	1.30b	14.14b	31.75b

Means in a column followed by a common letter are not significantly different at the 1 % level by DMRT.

Table 2: Vegetable oils controlling storage pests

Oil	Storage Pest	Author (S)
Castor	C. chinensis	Sangappa, 1977
Coconut	C. maculatus	Varma and Pandey, 1978
Cotton seed	C. chinensis	Sangappa, 1977
	C. maculatus	Pandey, et al., 1981
	S. oryzae	de Oca et al., 1978
	S. granarius	Yun-Tai Qi and Burkholder, 1981
Groundnut	S. cerealella	de Oca et al., 1978
	C. maculatus	IITA, 1976

	S. granarius	Varma and Pandey, 1978
		Yun-Tai Qi and Burkholder, 1981
Maize	C. maculatus	Akelo-Tsegah, 1976
		Singh, et al., 1978
	A. obtectus	Magoya et al., 1982
		van Rheenenet al., in press
	C. chinensis	Cruz and Cardona, 1981
	S. granarius	Yun-Tai Qi and Burkholder, 1981
	S. oryzae	de Oca et al., 1978
	S. cerealella	de Oca et al., 1978
Mustard	C. chinensis	Sangappa, 1977
	C. maculatus	Varma and Pandey, 1978
Neem	C. chinensis	Pandey et al., 1976
		Sangappa, 1977
Palm	S. oryzae	de Oca et al., 1978
	S. cerealella	de Oca et al., 1978
Rice	C. maculatus	Pandey et al., 1981
Sunflower	C. chinensis	Sangappa, 1977
	C. maculatus	Pandey et al., 1981
Sesame	C. maculatus	Varma and Pandey, 1978
Soybean	C. chinensis	Cruz and Cardona, 1981
	S. granarius	Yun-Tai Qi and Burkbolder, 1981
	S. oryzae	de Oca et al., 1978
	S cerealella	de Oca et al., 1978
Sunflower	A obtectus	Magoya et al., 1982
		van Rheenen et al., in press
	C. chinensis	Sangppa, 1977
	C maculatus	Varma and Pandey, 1978
Paraffin	C. maculatus	Calderon, 1979

Table 3 : Storage pests controlled by vegetable oils

Oil	Storage Pest	Author (S)
Cajanus cajan	Callosobruchus chinensis (L.)	Sangappa, 1977
Cicer arietinum	C. chinensis (L.)	Pandey et al., 1976
	C. maculatus (F.)	Calderon, 1979
Phaseolus vulgaris	A. obtectus (Say)	Magoya et al., 1982
		van Theenen et al., in press
	Zabrotes subfasciatus (Boh.)	Hill and van Schooven, 1981
		van Schooven, 1978

Sorghum vu/gare	Sitotroga cerealella (Ol.)	de Oca et al., 1978
	S. oryzae (L.)	de Oca et al., 1978
Triticum vu/gare	S. cerealella	de Oca et al., 1978
	S. granarius (1.)	Yun-Tai Qi and Burkholder, 1981
Vigna radiate	S. oryzae (L.)	de Oca et al., 1978
	C. maculatus (F.)	Pandey et al., 1981
		Varma and Pandey, 1978
Vigna unguiculata	C. chinensis (L.)	Cruz and Cardona, 1981
	C. maculatus (F.)	Akelo-Tsegah, 1976
		IITA, 1976
		Singh et al., 1978
Zea mays	S. cerealella (Ol.)	de Oca et al., 1978
	S. oryzae (L.)	de Oca et al 1978

Prospects of Insecticides from Plants

Acceptance of botanical insecticides for the control of storage insect pests by small scale farmers is influenced by the following parameters: availability, safety, quality and cost. Application of the control methods is simple and does not need sophisticated equipment add to the desirability of the method.

The search for more plant species with insecticidal properties is being pursued by scientists from all over the world. However, a systematic evaluation and identification of chemical and physical characteristics of active constituents should be accelerated. It is on this basis where knowledge on their efficacy in time, various levels of dosages and knowledge on toxicity on target pests will be valuable in field application.

Entomopathogens for the Control of Storage Pests

Several disease pathogens from insect pests of stored products have been studied. These diseases are caused by pathogenic bacteria, viruses, fungi, nematodes and protozoa and are frequently fatal to these pests especially during the larval stage. The young larvae are the most sensitive to the pathogens.

Insect pathogens can be isolated from infected insects and then cultured in the laboratory except for the viruses and protozoa which can only be cultured in living insects. Dust preparations or aqueous suspension of these pathogens can be applied to stored products in bulk in much the same way as conventional insecticides.

Another way of application of the pathogens is through food baiting with peromone or other attractants. The baited food may be placed close to the stored products. This method was effective for dissemination of pathogenic protozoa to kill beetles and also viral and bacterial pathogens of moths. Shapes et al., (1977) suggested, however, that to be very effective, the contaminated insects must leave the source of attractant and spread the pathogen within the pest population. This could be achieved by using a pheromone that fades out quickly and where there is little or no shelter so that the infected pesta are inclined to move out after feeding.

In this method, transmission of the pathogens may occur in one of these ways: 1) larval eating on cadavers on infected larvae or adults, 2) consumption of infected stored food, 3) contamination during mating, and 4) infection from the female during oviposition to its progeny.

General Considerations

In the considering the potential use of microbial insecticides, it may be generalized that the mode of infection or mode of action of entomogahhogens or their toxic by-products may be divided into two groups; according to their natural port of entry into their hosts.

The first group includes bacteria, protozoa and viruses. These pathogens must be ingested in order to cause infection and later on mortality. Viruses are quite specific in their sites of development and multiply only in certain tissues within the body of their host. Bacteria may multiply throughout the tissues and body fluids of the host causing septecemia. The crystalliferous bacteria (Bacillus thuringiensis) may kill their hosts purely on the basis of the activity of their associated toxins.

The second group includes the fungi and nematodes which enter their host through their integument. These microorganisms are more subject to regulations by the environment. If the environment favours them, they multiply tremendously and easily colonize their hosts. Physical factors like temperature and humidity affect their survival and/or ability to cause infection. These will also affect the progress of infection and the susceptibility or resustance of the host.

Bacteria

Bacillus thuringiensis Berliner, the most common bacteria used for the control of stored insect pests, was isolated from diseased larvae of the Mediterranean flour moth, Ephestia knehniella Zeller in 1911.

It has been tested on a wide range of storage moths: Plodia interpunctella Hubner, E. elutella, E. cautella, S. cerealeela (Oliver) and Corcyra cephalonica (Stainton). It was effective against all the above species except S. cerealella. In this species the larva spends its life cycle within the single grain and hence probably does not have sufficient contact with the pathogen when applied on the surfaces (McGaughey, 1976).

In bulk wheat and corn infested by P. interpunctella and E. cautella, it has been demonstrated by McGaughey (1976, 1978) that good control (at least 92%), may be obtained if only the surface layers of the bulk are treated with dust or aqueous suspension of B. thuringiensis. The recommended depth of treatment was 100 mm at least and the pathogen had to be well-mixed to ensure an even treatment within this layer. The recommended dose rate for this particular preparation was 125 mg/kg. Admixture to surface layers of bulk grain during grain transfer and after the silo have been filled were found eaually effective. Viability of B. thuringiensis is slightly reduced after one year storage.

Viruses

A nuclear polyhedrosis and a granulosis viruses have been isolated from E. cautella. Both severely infect P. interpunctella; whereas granulosis virus from P. interpunctella does not cross-infect E cautella (Hunter, 1973). Another nuclear polyhedrosis virus has been isolated from C. cephalonica but it is not known whether it will cross-infect other moths. In all cases the young larvae are the most susceptible stage.

Granulosis virus controlled effectively moth on bulk wheat and corn when surface layer of bulk was treated with aqueous suspension or dust of the virus to a depth of at least 100 mm. The suspension or dust both contained 3.2×10^3 virus capsules/mg and the grain was treated at a rate of approximately 1.9 mg/kg.

The application of an aqueous suspension of the granulosis virus of P. interpunctella has also been found to be an effective protectant for stored in-shell almonds (Hunter, et al, 1973), and stored raisins (Hunter et al, 1979). High storage temperature, however, reduces the viability of the virus and hence lowers the efficiency of this method of control. At 27°C, the control, after 5 & 6 months, remained high at 93% and 77%, respectively. In contrast, at 32°C, control after four months was 87% but this dropped to 34% and then 16% after 5 & 6 months.

Protozoa

Several protozoa are known to be severe pathogens of the Coleopteran. The following protozoans were tested: Nosema whitei for Tribolium castaneum (Herbs") and T. confusum J. du Val, N. whitei and N. oryzaephili for Oryzaephilus surinamensis (Linn) and Mattesia trogoderma for Trogoderma spp. Another species, Nosema plodiae Kellen and Lindegreem, has been isolated from the moth Q interpunctella and its mode of transmission investigated. Its use in control has not been studied.

The admixture of protozoa spores in particular, N. whitei to stored grain has been discussed by Burges (1973). He suggests that it would be necessary to admix a pathogen cocktail to control the range of Coleoptera that might be encountered in a particular storage situation.

Fungi

There are many fungi (hyphomycetous species) which grow on insects. Some are saprophytic while others are parasitic. Those most widely encountered are members of the genera Beauveria, Metarhizium and Isaria, and to a lesser extent Aspergillus, Cephalosporium, Sorosporella, and Hirsutella.

The commonest species is Metarhizium anisopliae (Metch) (Sor., the cause of the green muscardine disease of river' insects). This fungus has been seen to penetrate the integuments of several insects. In a recent study by Quiniones (1986) on the control of stored pests of copra, particularly the copra beetle, diluted spores of M. anisopliae up to 4.8 x 10-6 when topically brushed on the surface of stored copra, was effective to control this beetle.

Insect Resistance : Its Impact on Microbial Control of Insect Pests

Insects are among the earliest and most successful group of animals that exist in a myriad of environment where the potential for infection by microorganisms and parasites is great. Insects have also demonstrated considerable ability to develop resistance to conventional insecticides; and more than 500 species have developed resistance to one or more chemicals. As part of a survival strategy, insects have evolved numerous and effective resistance and defence mechanism to most of the conventional chemical insecticides with possession of genes for high levels of oxidase, esterases, glutathione-s-transferases, "insensitive" acetylcholinesterase (AChE), and nerve insensitivity to pyrethroids. Similarly can an insect species susceptible to a pathogen become resistant to more microorganism ?. Host specificity observed with many insect pathogens demonstrates that insect species are naturally resistant to these microorganisms. Indeed, insects that are susceptible to a pathogen can show resistance to various entomopathogens and try to resist infection through morphological, behavioural, developmental (like maturation immunity), physiological, nutritional, biochemical and molecular genetic mechanism etc.

In this review different aspects of insects immunity, viz. passive and active defence mechanism against foreign invaders in comparison with vertebrate immunity has been presented. How certain parasites and pathogens like fungus, bacteria,protozoa, nematodes and virus of insect pests evolved strategies for avoiding both the external barrier as well as the internal immune defence posed by the host insects has also been discussed. Understanding of current knowledge of insect haematology, and molecular basis of the insect biochemical and cellular defence has been stressed for the proper management of pests especially by using various biocontrol agents like parasites and pathogens, including Bt transgenic plants and also for the control of certain insect vectors which carries pathogens that cause human diseases by way of transforming insects themselves.

Insects vs. Vertebrate Immunity

Compared to vertebrates, insects do not possess the ability to produce antibodies (immunoglobulins) and do not use immunoglobulin as recognition molecules in the classical sense, against foreign antigen and hence antigenic memory appears to be lacking i.e. (non-memory type). Further, they do not produce alpha/beta interferons in response to viral infections.

Nevertheless, they are capable of "immune" reactions which appears to be predominantly cellular in nature. Several haemolymph induced antibacterial proteins have been reported to be broad-spectrum in their action which are produced in insects in response to bacterial challenger and of shorter duration in nature. This would suggest that analogy to the phenomenon of immunity in vertebrate may be inappropriate, and hence immunity in insects is different from immunity in vertebrates. The defence mechanism in insect is broadly classified into two broad groups. The first one is non-specific immunity which consists of structural and passive barriers like cuticle, gut physio-chemical properties, and peritrophic membrane. The second one is specific immune system involving cellular and humoral immunity which includes the activation of phenoloxoidase cascade, phagocytosis, nodulation (haemocyte aggregation) and encapsulation especially with reference to, bateria,fungi, protozoa including nematode invaders.

Non-specific Defence Mechanisms

Morphological

An insect resists entomopathogenic nematode (EPN) belonging to Steinernema and Heterorhabditis infection through behavioural, physical, or physiological means, Unlike vertebrates which have extensive exposure of epithelial cells to the external environment, insects have extensive protection of their epithelial tissue. The chitinous cuticle of the insect covers virtually all external surface, even extending through the foregut, hindgut and tracheal tubes, constituting the first line of passive defence in insects. Unlike that of the fore and hindgut, the epithelium of the insect midgut does not have a cuticular lining. However, in many insects there is a membrane called the "peritrophic membrane" which apparently functions to protect cells of the midgut from injury from hard (or) sharp particles of food performing much the same function as mucus in the mammalian alimentary tract. Physical resistance occurs when the nematode cannot penetrate the integument or the cocoon of a host insect. *Romanomermis culicivorax* has difficulty in penetrating the integument of older mosquito larvae. Dauer juveniles of *Steinernema carpocapsae* cannot penetrate the silken cocoons of hymenopteran parasitoids. Spiracular openings are portals of entry for EPN, but sieve plates over the spiracles, especially with scarab larvae, may deny nematodes access through this entry point. Avoidance of nematodes due to the presence of thick peritrophic membrane can act as a morphological defences against EPN.

Behavioural

Behavioural resistance occurs when the insect actively avoids or repels the nematode. Extremely active mosquito species had a lower prevalence of infection by the mermithid *R.. culicivorax* than less active ones. Scarab larvae may avoid infection by wiping nematodes away from the mouth. Younger instars of black fly larvae are resistant to infection by *Steinernema carpocapsae* because the comparatively large nematode is excluded from the insect's mouth. Aggressive behaviour (grooming with legs, mouth parts, and rasker) of *Popillia japonica* larvae when nematodes are present on the cuticle thereby removing and/or killing nematodes on the cuticle has also been reported.

Physiological

Under defence mechanisms, the high gut pH, presence of protease etc; can be found to be detrimental to the infective juveniles of *H. bacteriophora*. Similarly low gut pH, absence of protease etc; in insect system though not play a key or important resistant mechanism to bacterial pathogen like *Bacillus thuringiensis* (Bt) and Baculoviruses comprising nucleopolyhedrovirus (NPV) and granuloviruses (GV); have a role to play for their low susceptibility to bacterial and viral pathogens. Grasshoppers/locusts elevate their body temperatures higher than ambient via, habitat selection and/or orientation to solar radiations called 'basking' by way intercepting the solar radiation and raising internal thoracic temperature ranging from 380 to 420 c thereby showing 'behavioural fever' response.

Such a body temperature is predicted to inhibit fungal proliferation, thus giving the host immune system an edge in suppressing the fungus germination and growth which normally lakes place from 25 ° to 30 ° c, thereby reducing the infection. Thus, thermoregulation by grasshoppers, *Melanoplus sanguinipes* has been shown to reduce mycosis caused by *Beauveria bassiana* and *Metarhizium anisopliae* (Ouedraogo *et al.*,2003). Physiological resistance to infection involves the destruction of the nematode by digestive enzymes in the insect's alimentary tract and the melanization and encapsulation of the nematode within the hemocoel. Melanotic encapsulation of *Steinernema carpocapsae* has been reported in larvae of several mosquito species. Although the nematode is encapsulated, the majority of the larvae die of septicaemia caused by the bacterium *Xenorhabdus nematophilus*, which is mutualistically associated with this nematode.

Biochemical

Eicosanoids is a collective term for all biologically active, oxygenated metabolites of 20: 4n-6 and two other czo poly unsaturated fatty acids. Given the importance of eicosanoid signalling to the insect immune system, interference with eicosanoid metabolism would seem to be a sensible strategy, and thus a potentially important virulence factor for an entomopathogen. Mediation of bacterial clearance has been reported in insects by eicosanoids by way of nodulation on *Manduca sexta, Agrotis ipsilon, Pseudaletia unipuncta* and *Bombyx mori etc.*

Biotechnological (Molecular)

Very little is known concerning insect defence against virus infection although insect haemocytes can provide cell-mediated immunity to bacterial pathogens through phagocytes and encapsulation. Neither cell mediated or humoral immunity has been demonstrated against virus infection in insects. "Apoptosis" – distinctive type of programmed cell death – a phenomenon evolved as a primitive viral defence in certain vertebrate animals and invertebrates lacking humoral immunity to function as antiviral defence mechanism is gaining importance in cellular defence against viral infections (Narayanan, 1997). However, insect baculoviruses like nuclear polyhedrosis virus, granulosis virus and other DNA viruses, of insects evolved methods apparently to bypass this defence phenomenon of Apoptosis by directly blocking this response with possession of *p35* gene. The above findings leads to the future possibility of blocking the apoptosis for increasing virulence of certain baculoviruses as well as for the determination of host range of certain baculovirus and development of robust cells for in vitro multiplication of insect viruses and development of genetically improved parasitoids.

Genetic

Genetic resistance to bacterial and viral (Briese 1986b;) pathogens has been recorded against pestiferous insects. High levels of resistance to the δ – endotoxin of *B. thuringiensis* subspecies *kurstaki* have been recorded for the Indian meal moth, *Plodia interpunctella*, a pest of stored-grain and cereal products. The resistant trait is incompletely autosomal, recessive, and several alleles or genes are believed to be involved and resistance in this insect is linked to an alteration in toxin-membrane binding of the midgut cells.

Resistance of the silkworm to viral diseases such as nuclear (NPV) and cytoplasmic polyhedrosis virus (CPV) and infections flacherie virus (IFV) is controlled by polygenes. The polygenes are supposed to be mainly concerned with defence mechanism of midgut such as antiviral activity of gut juice, characteristic of peritropic membrane, etc. On the other hand non –susceptibility to densonucleosis is controlled by recessive (*nsd* – 1, *nsd* – 2) or dominant (*nsd* –1) major genes. The major gene may cause a deficiency of an enzyme involved in viral multiplication or in the receptor synthesis within the midgut cell. Polygenic resistance can be introgressed into a silkworm variety by selection in a breeding programme of the silkworm variety. A hybrid of two strains usually shows high heterosis in polygenic resistance to viral disease. The breeding procedure for non-susceptible variety to DNV is much easier because the mechanism of non-susceptibility is controlled by a single major gene. The gene can be introduced into the breeding programme or transferred to an existing superior variety by back crossing

Specific Defence Mechanisms/Immunity

Cellular Immunity

In a cellular defence mechanism, unlike vertebrate which has red blood corpuscles (RBC) and white blood corpuscles (WBC) in a closed circulatory system, insects with open body cavity lack lymphocytes, the major source of vertebrate immunity to virus infection. But they have only free blood cells called haemocytes. Different types of blood cells have important roles in the protection of insects against invading microorganisms. Hence identification and classification of various types of insect blood cells based on the structure and function is important. Among the six major group of insect haemocytes in recognizing the "self" (Isografts) and non-self (Allografts), plasmocytes and granulocytes, are the major effector cells and they react to foreign invaders either by phagocytosing like microorganisms or nodulating and encapsulating the objects too large to be individually engulfed,viz. metazoan parasite by way of haemocytes attaching and forming many layers which often become melanotic, thereby causing the death of the parasitoid through starvation and or anoxia mechanism. Changes in total haemocytes (THC) during growth and development of healthy insects have been reported by a number of workers (Narayanan1976). Drastic reduction in number of haemocytes during various microbial infection has also been reported by several workers. Infection by *B. bassiana* results in a gradual suppression of the phagocytic competence of circulating haemocytes and alteration in both total and differential haemocyte counts (DHC) has been reported in the case of fungal, bacterial, viral (Narayanan, 1979), and parasitic infection.

Humoral Immunity

Humoral reactions require several hours for their full expression and involve induced synthesis of antibacterial proteins, including 'cecropins', 'attacins', 'diptericins' and 'defensins'. The detergent properties of their antibacterial proteins disrupt cell membranes of the invading bacteria. Insects also synthesizing lysozome which enzymatically attack bacteria by hydrolyzing their peptidoglycan cell walls. Haemocytic responses feature direct interactions between circulating haemocyte and bacteria, these typically take place immediately after infections.Humoral reaction involve synthesis and release of several antibacterial (immuno) proteins. The antibacterial nature of gut contents and partial characterization of haemolymph bacterial proteins has been reported.

There are several families of characterized antibacterial proteins like cecropins, attacins, diptericins and defensins including recent identification of "Hemolin" (Sun et al., 1990) which belongs to the immunoglobulin superfamily. These insect antibacterial proteins are the best characterized invertebrate antibacterial factors and they have counterparts in mammals. The role of these proteins in non-self recognition as well as acting both prokaryotic and eukaryotic cells including human infectious parasites has been well studied.

4

DNA Vaccination

DNA vaccination is a technique for protecting an organism against disease by injecting it with genetically engineered DNA to produce an immunological response. Nucleic acid vaccines are still experimental, and have been applied to a number of viral, bacterial and parasitic models of disease, as well as to several tumour models. DNA vaccines have a number of advantages over conventional vaccines, including the ability to induce a wider range of immune response types.

Vaccines are among the greatest achievements of modern medicine – in industrial nations, they have eliminated naturally-occurring cases of smallpox, and nearly eliminated polio, while other diseases, such as typhus, rotavirus, hepatitis A and B and others are well controlled. Conventional vaccines, however, only cover a small number of diseases, and infections that lack effective vaccines kill millions of people every year, with AIDS, hepatitis C and malaria being particularly common.

First generation vaccines are whole-organism vaccines – either live and weakened, or killed forms. Live, attenuated vaccines, such as smallpox and polio vaccines, are able to induce killer T-cell (T_C or CTL) responses, helper T-cell (T_H) responses and antibody immunity. However, there is a small risk that attenuated forms of a pathogen can revert to a dangerous form, and may still be able to cause disease in immunocompromised people (such as those with AIDS). While killed vaccines do not have this risk, they cannot generate specific killer T cell responses, and may not work at all for some diseases. In order to minimise these risks, so-called *second generation vaccines* were developed. These are subunit vaccines, consisting of defined protein antigens (such as tetanus or diphtheria toxoid) or recombinant protein components (such as the hepatitis B surface antigen). These, too, are able to generate T_H and antibody responses, but not killer T cell responses. DNA vaccines are *third generation vaccines*, and are made up of a small, circular piece of bacterial DNA (called a plasmid) that has been genetically engineered to produce one or two specific proteins (antigens) from a pathogen. The vaccine DNA is injected into the cells of the body, where the "inner machinery" of the host cells "reads" the DNA and converts it into pathogenic proteins. Because these proteins are recognised as foreign, when they are processed by the host cells and displayed on their surface, the immune system is alerted, which then triggers a range of immune responses. These DNA vaccines developed from "failed" gene therapy experiments. The first demonstration of a plasmid-induced immune response was when mice inoculated with a plasmid expressing human growth hormone elicited antibodies instead of altering growth.

Current Use

Thus far, few experimental trials have evoked a response sufficiently strong enough to protect against disease, and the usefulness of the technique, while tantalizing, remains to be conclusively proven in human trials. However, in June 2006 positive results were announced for a bird flu DNA vaccine and a veterinary DNA vaccine to protect horses from West Nile virus has been approved. In August 2007, a preliminary study in DNA vaccination against multiple sclerosis was reported as being effective.

Plasmid Vectors for Use in Vaccination

Vector Design

DNA vaccines elicit the best immune response when highly active expression vectors are used. These are plasmids which usually consist of a strong viral promoter to drive the in vivo transcription and translation of the gene (or complementary DNA) of interest. Intron A may sometimes be included to improve mRNA stability and hence increase protein expression. Plasmids also include a strong polyadenylation/transcriptional termination signal, such as bovine growth hormone or rabbit beta-globulin polyadenylation sequences. Multicistronic vectors are sometimes constructed to express more than one immunogen, or to express an immunogen and an immunostimulatory protein. Because the plasmid is the "vehicle" from which the immunogen is expressed, optimising vector design for maximal protein expression is essential. One way of enhancing protein expression is by optimising the codon usage of pathogenic mRNAs for eukaryotic cells. Pathogens often have different AT contents than the species being immunized, so altering the gene sequence of the immunogen to reflect the codons more commonly used in the target species may improve its expression.

Another consideration is the choice of promoter. The SV40 promoter was conventionally used until research showed that vectors driven by the Rous Sarcoma Virus (RSV) promoter had much higher expression rates. More recently, expression rates have been further increased by the use of the cytomegalovirus (CMV) immediate early promoter. Inclusion of the Mason-Pfizer monkey virus (MPV)-CTE with/without rev increased envelope expression. Furthermore the CTE+rev construct was significantly more immunogenic then CTE alone vector. Additional modifications to improve expression rates have included the insertion of enhancer sequences, synthetic introns, adenovirus tripartite leader (TPL) sequences and modifications to the polyadenylation and transcriptional termination sequences. An example of DNA vaccine plasmid is pVAC, it uses SV40 promoter.

Vaccine Insert Design

Immunogens can be targeted to various cellular compartments in order to improve antibody or cytotoxic T-cell responses. Secreted or plasma membrane-bound antigens are more effective at inducing antibody responses than cytosolic antigens, while cytotoxic T-cell responses can be improved by targeting antigens for cytoplasmic degradation and subsequent entry into the major histocompatibility complex (MHC) class I pathway. This is usually accomplished by the addition of N-terminal ubiquitin signals.

The conformation of the protein can also have an effect on antibody responses, with "ordered" structures (like viral particles) being more effective than unordered structures. Strings of minigenes (or MHC class I epitopes) from different pathogens are able to raise cytotoxic T-cell responses to a number of pathogens, especially if a TH epitope is also included.

Delivery Methods

DNA vaccines have been introduced into animal tissues by a number of different methods. The two most popular approaches are injection of DNA in saline, using a standard hypodermic needle, and gene gun delivery. A schematic outline of the construction of a DNA vaccine plasmid and its subsequent delivery by these two methods into a host is illustrated at Scientific American. Injection in saline is normally conducted intramuscularly (IM) in skeletal muscle, or intradermally (ID), with DNA being delivered to the extracellular spaces. This can be assisted by electroporation; by temporarily damaging muscle fibres with myotoxins such as bupivacaine; or by using hypertonic solutions of saline or sucrose. Immune responses to this method of delivery can be affected by many factors, including needle type, needle alignment, speed of injection, volume of injection, muscle type, and age, sex and physiological condition of the animal being injected.

Gene gun delivery, the other commonly used method of delivery, ballistically accelerates plasmid DNA (pDNA) that has been adsorbed onto gold or tungsten microparticles into the target cells, using compressed helium as an accelerant.

Alternative delivery methods have included aerosol instillation of naked DNA on mucosal surfaces, such as the nasal and lung mucosa, and topical administration of pDNA to the eye and vaginal mucosa. Mucosal surface delivery has also been achieved using cationic liposome-DNA preparations, biodegradable microspheres, attenuated *Shigella* or *Listeria* vectors for oral administration to the intestinal mucosa, and recombinant adenovirus vectors.

The method of delivery determines the dose of DNA required to raise an effective immune response. Saline injections require variable amounts of DNA, from 10 µg-1 mg, whereas gene gun deliveries require 100 to 1000 times less DNA than intramuscular saline injection to raise an effective immune response. Generally, 0.2 µg – 20 µg are required, although quantities as low as 16 ng have been reported. These quantities vary from species to species, with mice, for example, requiring approximately 10 times less DNA than primates. Saline injections require more DNA because the DNA is delivered to the extracellular spaces of the target tissue (normally muscle), where it has to overcome physical barriers (such as the basal lamina and large amounts of connective tissue, to mention a few) before it is taken up by the cells, while gene gun deliveries bombard DNA directly into the cells, resulting in less "wastage".

Another approach to DNA vaccination is expression library immunization (ELI). Using this technique, potentially all the genes from a pathogen can be delivered at one time, which may be useful for pathogens which are difficult to attenuate or culture. ELI can be used to identify which of the pathogen's genes induce a protective response. This has been tested with *Mycoplasma pulmonis*, a murine lung pathogen with a relatively small genome, and it was found that even partial expression libraries can induce protection from subsequent challenge.

Immune Response Raised by DNA Vaccines

Helper T-Cell Responses

Antigen presentation stimulates T cells to become either "cytotoxic" CD8+ cells or "helper" CD4+ cells. Cytotoxic cells directly attack other cells carrying certain foreign or abnormal molecules on their surfaces. Helper T cells, or Th cells, coordinate immune responses by communicating with other cells. In most cases, T cells only recognize an antigen if it is carried on the surface of a cell by one of the body's own MHC, or major histocompatibility complex, molecules.

DNA immunization is able to raise a range of T_H responses, including lymphoproliferation and the generation of a variety of cytokine profiles. A major advantage of DNA vaccines is the ease with which they can be manipulated to bias the type of T-cell help towards a TH1 or TH2 response. Each type of response has distinctive patterns of lymphokine and chemokine expression, specific types of immunoglobulins expressed, patterns of lymphocyte trafficking, and types of innate immune responses generated.

Raising of Different Types of T-cell Help

The type of T-cell help raised is influenced by the method of delivery and the type of immunogen expressed, as well as the targeting of different lymphoid compartments. Generally, saline needle injections (either IM or ID) tend to induce TH1 responses, while gene gun delivery raises TH2 responses.

This is true for intracellular and plasma membrane-bound antigens, but not for secreted antigens, which seem to generate TH2 responses, regardless of the method of delivery.

Generally the type of T-cell help raised is stable over time, and does not change when challenged or after subsequent immunizations which would normally have raised the opposite type of response in a naïve animal. However, Mor *et al.*. (1995) immunized and boosted mice with pDNA encoding the circumsporozoite protein of the mouse malarial parasite *Plasmodium yoelii* (PyCSP) and found that the initial TH2 response changed, after boosting, to a TH1 response.

Mechanistic Basis for Different Types of T-Cell Help

It is not understood how these different methods of DNA immunization, or the forms of antigen expressed, raise different profiles of T-cell help. It was thought that the relatively large amounts of DNA used in IM injection were responsible for the induction of TH1 responses. However, evidence has shown no differences in TH type due to dose. It has been postulated that the type of T-cell help raised is determined by the differentiated state of antigen presenting cells. Dendritic cells can differentiate to secrete IL-12 (which supports TH1 cell development) or IL-4 (which supports TH2 responses). pDNA injected by needle is endocytosed into the dendritic cell, which is then stimulated to differentiate for TH1 cytokine production, while the gene gun bombards the DNA directly into the cell, thus bypassing TH1 stimulation.

Practical Uses of Polarised T-Cell Help

This polarisation in T-cell help is useful in influencing allergic responses and autoimmune diseases. In autoimmune diseases, the goal would be to shift the self-destructive TH1 response (with its associated cytotoxic T cell activity) to a non-destructive TH2 response. This has been successfully applied in predisease priming for the desired type of response in preclinical models and somewhat successful in shifting the response for an already established disease.

Cytotoxic T-cell Responses

One of the greatest advantages of DNA vaccines is that they are able to induce cytotoxic T lymphocytes (CTL) without the inherent risk associated with live vaccines. CTL responses can be raised against immunodominant and immunorecessive CTL epitopes, as well as subdominant CTL epitopes, in a manner which appears to mimic natural infection. This may prove to be a useful tool in assessing CTL epitopes of an antigen, and their role in providing immunity.

Cytotoxic T-cells recognise small peptides (8-10 amino acids) complexed to MHC class I molecules (Restifo et al., 1995). These peptides are derived from endogenous cytosolic proteins which are degraded and delivered to the nascent MHC class I molecule within the endoplasmic reticulum (ER). Targeting gene products directly to the ER (by the addition of an amino-terminal insertion sequence) should thus enhance CTL responses. This has been successfully demonstrated using recombinant vaccinia viruses expressing influenza proteins, but the principle should be applicable to DNA vaccines too.

Targeting antigens for intracellular degradation (and thus entry into the MHC class I pathway) by the addition of ubiquitin signal sequences, or mutation of other signal sequences, has also been shown to be effective at increasing CTL responses.

CTL responses can also be enhanced by co-inoculation with co-stimulatory molecules such as B7-1 or B7-2 for DNA vaccines against influenza nucleoprotein, or GM-CSF for DNA vaccines against the murine malaria model *P. yoelii*. Co-inoculation with plasmids encoding co-stimulatory molecules IL-12 and TCA3 have also been shown to increase CTL activity against HIV-1 and influenza nucleoprotein antigens.

Humoral (Antibody) Response

Antibody responses elicited by DNA vaccinations are influenced by a number of variables, including type of antigen encoded; location of expressed antigen (i.e. intracellular vs. secreted); number, frequency and dose of immunizations; site and method of antigen delivery, to name a few.

Kinetics of Antibody Response

Humoral responses after a single DNA injection can be much longer-lived than after a single injection with a recombinant protein. Antibody responses against hepatitis B virus (HBV) envelope protein (HBsAg) have been sustained for up to 74 weeks without boost, while life-long maintenance of protective response to influenza haemagglutinin has been demonstrated in mice after gene gun delivery. Antibody-secreting cells migrate to the bone marrow and spleen for long-term antibody production, and are generally localised there after one year.

DNA-raised antibody responses rise much more slowly than when natural infection or recombinant protein immunization occurs. It can take as long as 12 weeks to reach peak titres in mice, although boosting can increase the rate of antibody production. This slow response is probably due to the low levels of antigen expressed over several weeks, which supports both primary and secondary phases of antibody response.

Additionally, the titres of specific antibodies raised by DNA vaccination are lower than those obtained after vaccination with a recombinant protein. However, DNA immunization-induced antibodies show greater affinity to native epitopes than recombinant protein-induced antibodies. In other words, DNA immunization induces a qualitatively superior response. Antibody can be induced after just one vaccination with DNA, whereas recombinant protein vaccinations generally require a boost. As mentioned previously, DNA immunization can be used to bias the TH profile of the immune response, and thus the antibody isotype, which is not possible with either natural infection or recombinant protein immunization. Antibody responses generated by DNA are useful not just in vaccination but as a preparative tool, too. For example, polyclonal and monoclonal antibodies can be generated for use as reagents.

Mechanistic Basis for DNA Raised Immune Responses

DNA Uptake Mechanism

When DNA uptake and subsequent expression was first demonstrated *in vivo* in muscle cells, it was thought that these cells were unique in this ability because of their extensive network of T-tubules. Using electron microscopy, it was proposed that DNA uptake was facilitated by caveolae (or, non-clathrin coated pits). However, subsequent research revealed that other cells (such as keratinocytes, fibroblasts and epithelial Langerhans cells) could also internalize DNA. This phenomenon has not been the subject of much research, so the actual mechanism of DNA uptake is not known.

Two theories are currently popular – that *in vivo* uptake of DNA occurs non-specifically, in a method similar to phago- or pinocytosis, or through specific receptors. These might include a 30kDa surface receptor, or macrophage scavenger receptors. The 30kDa surface receptor binds very specifically to 4500-bp genomic DNA fragments (which are then internalised) and is found on professional APCs and T-cells. Macrophage scavenger receptors bind to a variety of macromolecules, including polyribonucleotides, and are thus also candidates for DNA uptake. Receptor mediated DNA uptake could be facilitated by the presence of polyguanylate sequences. Further research into this mechanism might seem pointless, considering that gene gun delivery systems, cationic liposome packaging, and other delivery methods bypass this entry method, but understanding it might be useful in reducing costs (e.g. by reducing the requirement for cytofectins), which will be important in the food animals industry.

Antigen Presentation by Bone Marrow-derived Cells

Studies using chimeric mice have shown that antigen is presented by bone-marrow derived cells, which include dendritic cells, macrophages and specialised B-cells called professional antigen presenting cells (APC) Iwasaki et al., 1997). After gene gun inoculation to the skin, transfected Langerhans cells migrate to the draining lymph node to present antigen. After IM and ID injections, dendritic cells have also been found to present antigen in the draining lymph node and transfected macrophages have been found in the peripheral blood. Besides direct transfection of dendritic cells or macrophages, cross priming is also known to occur following IM, ID and gene gun deliveries of DNA. Cross priming occurs when a bone marrow-derived cell presents peptides from proteins synthesised in another cell in the context of MHC class 1. This can prime cytotoxic T-cell responses and seems to be important for a full primary immune response.

Role of the Target Site

IM and ID delivery of DNA initiate immune responses differently. In the skin, keratinocytes, fibroblasts and Langerhans cells take up and express antigen, and are responsible for inducing a primary antibody response. Transfected Langerhans cells migrate out of the skin (within 12 hours) to the draining lymph node where they prime secondary B- and T-cell responses. In skeletal muscle, on the other hand, striated muscle cells are most frequently transfected, but seem to be unimportant in mounting an immune response. Instead, IM inoculated DNA "washes" into the draining lymph node within minutes, where distal dendritic cells are transfected and then initiate an immune response. Transfected myocytes seem to act as a "reservoir" of antigen for trafficking professional APCs.

Maintenance of Immune Response

DNA vaccination generates an effective immune memory via the display of antigen-antibody complexes on follicular dendritic cells (FDC), which are potent B-cell stimulators. T-cells can be stimulated by similar, germinal centre dendritic cells. FDC are able to generate an immune memory because antibodies production "overlaps" long-term expression of antigen, allowing antigen-antibody immunocomplexes to form and be displayed by FDC.

Interferons

Both helper and cytotoxic T-cells can control viral infections by secreting interferons. Cytotoxic T cells usually kill virally infected cells. However, they can also be stimulated to secrete antiviral cytokines such as INF-γ and TNF-α, which don't kill the cell but place severe limitations on viral infection by down-regulating the expression of viral components. DNA vaccinations can thus be used to curb viral infections by non-destructive IFN-mediated control. This has been demonstrated for the hepatitis B virus. IFN-γ is also critically important in controlling malaria infections, and should be taken into consideration when developing anti-malarial DNA vaccines.

Modulation of the Immune Response

Cytokine Modulation

For a vaccine to be effective, it must induce an appropriate immune response for a given pathogen, and the ability of DNA vaccines to polarise T-cell help towards TH1 or TH2 profiles, and generate CTL and/or antibody when required, is a great advantage in this regard. This can be accomplished by modifications to the form of antigen expressed (i.e. intracellular vs. secreted), the method and route of delivery, and the dose of DNA delivered. However, it can also be accomplished by the co-administration of plasmid DNA encoding immune regulatory molecules, i.e. cytokines, lymphokines or co-stimulatory molecules. These "genetic adjuvants" can be administered a number of ways:

- as a mixture of 2 separate plasmids, one encoding the immunogen and the other encoding the cytokine;
- as a single bi- or polycistronic vector, separated by spacer regions; or
- as a plasmid-encoded chimera, or fusion protein.

In general, co-administration of pro-inflammatory agents (such as various interleukins, tumour necrosis factor, and GM-CSF) plus TH2 inducing cytokines increase antibody responses, whereas pro-inflammatory agents and TH1 inducing cytokines decrease humoral responses and increase cytotoxic responses (which is more important in viral protection, for example). Co-stimulatory molecules like B7-1, B7-2 and CD40L are also sometimes used.

This concept has been successfully applied in topical administration of pDNA encoding IL-10. Plasmid encoded B7-1 (a ligand on APCs) has successfully enhanced the immune response in anti-tumour models, and mixing plasmids encoding GM-CSF and the circumsporozoite protein of P. yoelii (PyCSP) has enhanced protection against subsequent challenge (whereas plasmid-encoded PyCSP alone did not). It was proposed that GM-CSF may cause dendritic cells to present antigen more efficiently, and enhance IL-2 production and TH cell activation, thus driving the increased immune response. This can be further enhanced by first priming with a pPyCSP and pGM-CSF mixture, and later boosting with a recombinant poxvirus expressing PyCSP. However, co-injection of plasmids encoding GM-CSF (or IFN-γ, or IL-2) and a fusion protein of *P. chabaudi* merozoite surface protein 1 (C-terminus)-hepatitis B virus surface protein (PcMSP1-HBs) actually abolished protection against challenge, compared to protection acquired by delivery of pPcMSP1-HBs alone.

The advantages of using genetic adjuvants are their low cost and simplicity of administration, as well as avoidance of unstable recombinant cytokines and potentially toxic, "conventional" adjuvants (such as alum, calcium phosphate, monophosphoryl lipid A, cholera toxin, cationic and mannan-coated liposomes, QS21, carboxymethylcellulose and ubenimix). However, the potential toxicity of prolonged cytokine expression has not been established, and in many commercially important animal species, cytokine genes still need to be identified and isolated. In addition, various plasmid encoded cytokines modulate the immune system differently according to the time of delivery. For example, some cytokine plasmid DNAs are best delivered after the immunogen pDNA, because pre- or co-delivery can actually decrease specific responses, and increase non-specific responses.

Immunostimulatory CpG Motifs

Plasmid DNA itself appears to have an adjuvant effect on the immune system. Bacterially derived DNA has been found to trigger innate immune defence mechanisms, the activation of dendritic cells, and the production of TH1 cytokines. This is due to recognition of certain CpG dinucleotide sequences which are immunostimulatory. CpG stimulatory (CpG-S) sequences occur twenty times more frequently in bacterially derived DNA than in eukaryotes. This is because eukaryotes exhibit "CpG suppression" – i.e. CpG dinucleotide pairs occur much less frequently than expected. Additionally, CpG-S sequences are hypomethylated. This occurs frequently in bacterial DNA, while CpG motifs occurring in eukaryotes are all methylated at the cytosine nucleotide. In contrast, nucleotide sequences which inhibit the activation of an immune response (termed CpG neutralising, or CpG-N) are over represented in eukaryotic genomes. The optimal immunostimulatory sequence has been found to be an unmethylated

CpG dinucleotide flanked by two 5' purines and two 3' pyrimidines. Additionally, flanking regions outside this immunostimulatory hexamer must be guanine-rich to ensure binding and uptake into target cells.

The innate system works synergistically with the adaptive immune system to mount a response against the DNA encoded protein. CpG-S sequences induce polyclonal B-cell activation and the upregulation of cytokine expression and secretion. Stimulated macrophages secrete IL-12, IL-18, TNF-α, IFN-α, IFN-β and IFN-γ, while stimulated B-cells secrete IL-6 and some IL-12.

Manipulation of CpG-S and CpG-N sequences in the plasmid backbone of DNA vaccines can ensure the success of the immune response to the encoded antigen, and drive the immune response toward a TH1 phenotype. This is useful if a pathogen requires a TH response for protection. CpG-S sequences have also been used as external adjuvants for both DNA and recombinant protein vaccination with variable success rates. Other organisms with hypomethylated CpG motifs have also demonstrated the stimulation of polyclonal B-cell expansion. However, the mechanism behind this may be more complicated than simple methylation – hypomethylated murine DNA has not been found to mount an immune response.

Most of the evidence for the existence of immunostimulatory CpG sequences comes from murine studies. Clearly, extrapolation of this data to other species should be done with caution – different species may require different flanking sequences, as binding specificities of scavenger receptors differ between species. Additionally, species such as ruminants may be insensitive to immunostimulatory sequences due to the large gastrointestinal load they exhibit. Further research may be useful in the optimisation of DNA vaccination, especially in the food animal industry.

Alternative Boosts

DNA-primed immune responses can be boosted by the administration of recombinant protein or recombinant poxviruses. "Prime-boost" strategies with recombinant protein have successfully increased both neutralising antibody titre, and antibody avidity and persistence, for weak immunogens, such as HIV-1 envelope protein. Recombinant virus boosts have been shown to be very efficient at boosting DNA-primed CTL responses. Priming with DNA focuses the immune response on the required immunogen, while boosting with the recombinant virus provides a larger amount of expressed antigen, leading to a large increase in specific CTL responses. Prime-boost strategies have been successful in inducing protection against malarial challenge in a number of studies.

Primed mice with plasmid DNA encoding *Plasmodium yoelii* circumsporozoite surface protein (PyCSP), then boosted with a recombinant vaccinia virus expressing the same protein had significantly higher levels of antibody, CTL activity and IFN-γ, and hence higher levels of protection, than mice immunized and boosted with plasmid DNA alone.

This can be further enhanced by priming with a mixture of plasmids encoding PyCSP and murine GM-CSF, before boosting with recombinant vaccinia virus. An effective prime-boost strategy for the simian malarial model *P. knowlesi* has also been demonstrated. Rhesus monkeys were primed with a multicomponent, multistage DNA vaccine encoding two liver-stage antigens-the circumsporozoite surface protein (PkCSP) and sporozoite surface protein 2 (PkSSP2)- and two blood stage antigens-the apical merozoite surface protein 1 (PkAMA1) and merozoite surface protein 1 (PkMSP1p42). They were then boosted with a recombinant canarypox virus encoding all four antigens (ALVAC-4). Immunized monkeys developed antibodies against sporozoites and infected erythrocytes, and IFN-γ-secreting T-cell responses against peptides from PkCSP. Partial protection against sporozoite challenge was achieved, and mean parasitemia was significantly reduced, compared to control monkeys. These models, while not ideal for extrapolation to *P. falciparum* in humans, will be important in pre-clinical trials.

Additional Methods of Enhancing DNA-Raised Immune Responses

Formulations of DNA

The efficiency of DNA immunization can be improved by stabilising DNA against degradation, and increasing the efficiency of delivery of DNA into antigen presenting cells. This has been demonstrated by coating biodegradable cationic microparticles (such as poly(lactide-co-glycolide) formulated with cetyltrimethylammonium bromide) with DNA. Such DNA-coated microparticles can be as effective at raising CTL as recombinant vaccinia viruses, especially when mixed with alum. Particles 300 nm in diameter appear to be most efficient for uptake by antigen presenting cells.

Alphavirus Vectors

Recombinant alphavirus-based vectors have also been used to improve DNA vaccination efficiency. The gene encoding the antigen of interest is inserted into the alphavirus replicon, replacing structural genes but leaving non-structural replicase genes intact. The Sindbis virus and Semliki Forest virus have been used to build recombinant alphavirus replicons. Unlike conventional DNA vaccinations, however, alphavirus vectors kill transfected cells, and are only transiently expressed. Also, alphavirus replicase genes are expressed in addition to the vaccine insert. It is not clear how alphavirus replicons raise an immune response, but it is thought that this may be due to the high levels of protein expressed by this vector, replicon-induced cytokine responses, or replicon-induced apoptosis leading to enhanced antigen uptake by dendritic cells.

Vector DNA

A vector is a DNA molecule used as a vehicle to transfer foreign genetic material into another cell. The four major types of vectors are plasmids, bacteriophages and other viruses, cosmids, and artificial chromosomes. Common to all engineered vectors are an origin of replication, a multicloning site, and a selectable marker. The vector itself is generally a DNA sequence that consists of an insert (transgene) and a larger sequence that serves as the "backbone" of the vector. The purpose of a vector which transfers genetic information to another cell is typically to isolate, multiply, or express the insert in the target cell. Vectors called expression vectors (expression constructs) specifically are for the expression of the transgene in the target cell, and generally have a promoter sequence that drives expression of the transgene. Simpler vectors called transcription vectors are only capable of being transcribed but not translated: they can be replicated in a target cell but not expressed, unlike expression vectors. Transcription vectors are used to amplify their insert. Insertion of a vector into the target cell is usually called transformation for bacterial cells, transfection for eukaryotic cells, although insertion of a viral vector is often called transduction.

Characteristics

Two common vectors are plasmids and viral vectors.

Plasmids

Plasmids are double-stranded generally circular DNA sequences that are capable of automatically replicating in a host cell. Plasmid vectors minimalistically consist of an origin of replication that allows for semi-independent replication of the plasmid in the host and also the transgene insert. Modern plasmids generally have many more features, notably including a "multiple cloning site" which includes nucleotide overhangs for insertion of an insert, and multiple restriction enzyme consensus sites to either side of the insert. In the case of plasmids utilized as transcription vectors, incubating bacteria with plasmids generates hundreds or thousands of copies of the vector within the bacteria in hours, and the vectors can be extracted from the bacteria, and the multiple cloning site can be cut by restriction enzymes to excise the hundredfold or thousandfold amplified insert. These plasmid transcription vectors characteristically lack crucial sequences that code for polyadenylation sequences and translation termination sequences in translated mRNAs, making protein expression from transcription vectors impossible. plasmids may be-conjugative/transmissible and non conjugative. conjugative-mediate DNA transfer through conjugation and therefore spread rapidly among the bacterial ceels of a population. e.g- F plasmid, many R and some col plasmids. nonconjugative- do not mediate DNA through conjugation. e.g.- many R and col plasmids.

Viral Vectors

Viral vectors are generally genetically-engineered viruses carrying modified viral DNA or RNA that has been rendered noninfectious, but still contain viral promoters and also the transgene, thus allowing for translation of the transgene through a viral promoter. However, because viral vectors frequently are lacking infectious sequences, they require helper viruses or packaging lines for large-scale transfection. Viral vectors are often designed for permanent incorporation of the insert into the host genome, and thus leave distinct genetic markers in the host genome after incorporating the transgene. For example, retroviruses leave a characteristic retroviral integration pattern after insertion that is detectable and indicates that the viral vector has incorporated into the host genome.

Transcription

Transcription is a necessary component in all vectors: the premise of a vector is to multiply the insert (although expression vectors later also drive the translation of the multiplied insert). Thus, even stable expression is determined by stable transcription, which generally depends on promoters in the vector. However, expression vectors have a variety of expression patterns: constitutive (consistent expression) or inducible (expression only under certain conditions or chemicals). This expression is based on different promoter activities, not post-transcriptional activities. Thus, these two different types of expression vectors depend on different types of promoters.

Viral promoters are often used for constitutive expression in plasmids and in viral vectors because they normally reliably force constant transcription in many cell lines and types.

Inducible expression depends on promoters that respond to the induction conditions: for example, the murine mammary tumour virus promoter only initiates transcription after dexamethasone application and the *Drosophilia* heat shock promoter only initiates after high temperatures. transcription is the synthesis of mRNA. Genetic information is copied from DNA to RNA.

Expression

Expression vectors require translation of the vector's insert, thus requiring more components than simpler transcription-only vectors. Expression vectors require sequences that encode for:

- Polyadenylation tail: Creates a polyadenylation tail at the end of the transcribed pre-mRNA that protects the mRNA from exonucleases and ensures transcriptional and translational termination: stabilizes mRNA production.

- Minimal UTR length: UTRs contain specific characteristics that may impede transcription or translation, and thus the shortest UTRs or none at all are encoded for in optimal expression vectors.

- Kozak sequence: Vectors should encode for a Kozak sequence in the mRNA, which assembles the ribosome for translation of the mRNA.

Features

Modern vectors may encompass additional features besides the transgene insert and a backbone:

- Promoter: Necessary component for all vectors: used to drive transcription of the vector's transgene.

- Genetic markers: Genetic markers for viral vectors allow for confirmation that the vector has integrated with the host genomic DNA.

- Antibiotic resistance: Vectors with antibiotic-resistance open reading frames allow for survival of cells that have taken up the vector in growth media containing antibiotics through antibiotic selection.

- Epitope: Vector contains a sequence for a specific epitope that is incorporated into the expressed protein. Allows for antibody identification of cells expressing the target protein.

- β-galactosidase: Some vectors contain a sequence for β-galactosidase, an enzyme that digests galactose, within which a multiple cloning site, the region in which a gene may be inserted, is located. An insert successfully ligated into the vector will disrupt the β-galactosidase gene and thus unable to digest galactose. Cells containing vector with an insert may be identified using blue/white selection by growing cells in media containing an analogue of galactose (X-gal). Cells expressing β-galactosidase (therefore doesn't contain an insert) appear as blue colonies. White colonies would be selected as those that may contain an insert. Other proteins which may function similarly as a reporter include green fluorescent protein and luciferase.

- Targeting sequence: Expression vectors may include encoding for a targeting sequence in the finished protein that directs the expressed protein to a specific organelle in the cell or specific location such as the periplasmic space of bacteria.

- Protein purification tags: Some expression vectors include proteins or peptide sequences that allows for easier purification of the expressed protein. Examples include polyhistidine-tag, glutathione-S-transferase, and maltose binding protein. Some of these tags may also allow for increased solubility of the target protein. The target protein is fused to the protein tag, but a protease cleavage site positioned in the polypeptide linker region between the protein and the tag allows the tag to be removed later.

HIV Vaccine

An HIV vaccine is the theoretical vaccine which would be given to persons without HIV in order to vaccinate them against getting HIV, the virus which causes AIDS. No effective vaccine against HIV exists. As there is not a known cure for AIDS, the search for a vaccine has become part of medical approaches against the disease.

It has been known for many years that HIV is an extremely difficult virus to render harmless, and no cure presently exists. Research into a vaccine is one of several strategies to reduce the worldwide harm from AIDS, with other approaches based upon antiviral treatments such as highly active antiretroviral therapy (HAART), and social approaches such as safe sex.

There is evidence that a vaccine may be possible. Work with monoclonal antibodies (MAb) has proven that the human body can defend itself against HIV, and certain individuals remain asymptomatic for decades after HIV infection. More recently in 2009, a number of potential candidates for antibodies and early stage results from clinical trials have been announced by various teams. However these are early results, and have either not been developed to the point of human testing, or not fully peer reviewed and replicated by other teams, at this time.

The urgency of the search for a vaccine against HIV stems from the AIDS-related death toll of over 25 million people since 1981. Indeed, in 2002, AIDS became the primary cause of mortality due to an infectious agent in Africa.

Alternative medical treatments to a vaccine do exist. Highly active antiretroviral therapy (HAART) has been highly beneficial to many HIV-infected individuals since its introduction in 1996 when the protease inhibitor-based HAART initially became available. HAART allows the stabilization of the patient's symptoms and viremia, but they do not cure the patient of HIV, nor of the symptoms of AIDS. And, importantly, HAART does nothing to prevent the spread of HIV through people with undiagnosed HIV infections. Safer sex measures have also proven insufficient to halt the spread of AIDS in the worst affected countries, despite some success in reducing infection rates. Therefore, an HIV vaccine is generally considered as the most likely, and perhaps the only way by which the AIDS pandemic can be halted. However, after over 20 years of research, HIV-1 remains a difficult target for a vaccine.

Difficulties in Developing an HIV Vaccine

In 1984, after the confirmation of the etiological agent of AIDS by scientists at the U.S. National Institutes of Health and the Pasteur Institute, the United States Health and Human Services Secretary Margaret Heckler declared that a vaccine would be available within two years. However, the classical vaccination approaches that have been successful in the control of various viral diseases by priming the adaptive immunity to recognize the viral envelope proteins have failed in the case of HIV-1. Some have stated that an HIV vaccine may not be possible without significant theoretical advances.

There are a number of factors that cause development of an HIV vaccine to differ from the development of other classic vaccines:

- Classic vaccines mimic natural immunity against reinfection generally seen in individuals recovered from infection; there are almost no recovered AIDS patients.

- Most vaccines protect against disease, not against infection; HIV infection may remain latent for long periods before causing AIDS.

- Most effective vaccines are whole-killed or live-attenuated organisms; killed HIV-1 does not retain antigenicity and the use of a live retrovirus vaccine raises safety issues.

- Most vaccines protect against infections that are infrequently encountered; HIV may be encountered daily by individuals at high risk.

- Most vaccines protect against infections through mucosal surfaces of the respiratory or gastrointestinal tract; the great majority of HIV infection is through the genital tract.

HIV Structure

The epitopes of the viral envelope are more variable than those of many other viruses. Furthermore, the functionally important epitopes of the gp 120 protein are masked by glycosylation, trimerisation and receptor-induced conformational changes making it difficult to block with neutralising antibodies. The ineffectiveness of previously developed vaccines primarily stems from two related factors.

- First, HIV is highly mutable. Because of the virus' ability to rapidly respond to selective pressures imposed by the immune system, the population of virus in an infected individual typically evolves so that it can evade the two major arms of the adaptive immune system; humoral (antibody-mediated) and cellular (mediated by T cells) immunity.

- Second, HIV isolates are themselves highly variable. HIV can be categorized into multiple clades and subtypes with a high degree of genetic divergence. Therefore, the immune responses raised by any vaccine need to be broad enough to account for this variability. Any vaccine that lacks this breadth is unlikely to be effective.

The difficulties in stimulating a reliable antibody response has led to the attempts to develop a vaccine that stimulates a response by cytotoxic T-lymphocytes.

Another response to the challenge has been to create a single peptide that contains the least variable components of all the known HIV strains.

Animal Model

The typical animal model for vaccine research is the monkey, often the macaque. Monkeys can be infected with SIV or the chimeric SHIV for research purposes. However, the well-proven route of trying to induce neutralizing antibodies by vaccination has stalled because of the great difficulty in stimulating antibodies that neutralise heterologous primary HIV isolates. Some vaccines based on the virus envelope have protected chimpanzees or macaques from homologous virus challenge, but in clinical trials, individuals who were immunised with similar constructs became infected after later exposure to HIV-1.

There are some differences between SIV and HIV that may introduce challenges in the use of an animal model.

As published on 27 November 2009 in Journal of Biology, there is new animal model strongly resembling that of HIV in humans. Generalized immune activation as a direct result of activated CD4+ T cell killing- performed in mice allows new ways of testing HIV behaviour.

Clinical Trials to Date

Several vaccine candidates are in varying phases of clinical trials.

Phase I

Most initial approaches have focused on the HIV envelope protein. At least thirteen different gp120 and gp160 envelope candidates have been evaluated, in the US predominantly through the AIDS Vaccine Evaluation Group. Most research focused on gp120 rather than gp41/gp160, as the latter are generally more difficult to produce and did not initially offer any clear advantage over gp120 forms. Overall, they have been safe and immunogenic in diverse populations, have induced neutralizing antibody in nearly 100% recipients, but rarely induced

CD8+ cytotoxic T lymphocytes (CTL). Mammalian derived envelope preparations have been better inducers of neutralizing antibody than candidates produced in yeast and bacteria. Although the vaccination process involved many repeated "booster" injections, it was very difficult to induce and maintain the high anti-gp120 antibody titers necessary to have any hope of neutralizing an HIV exposure.

The availability of several recombinant canarypox vectors has provided interesting results that may prove to be generalizable to other viral vectors. Increasing the complexity of the canarypox vectors by inclusion of more genes/epitopes has increased the percent of volunteers that have detectable CTL to a greater extent than did increasing the dose of the viral vector. Importantly, CTLs from volunteers were able to kill peripheral blood mononuclear cells infected with primary isolates of HIV, suggesting that induced CTLs could have biological significance. In addition, cells from at least some volunteers were able to kill cells infected with HIV from other clades, though the pattern of recognition was not uniform among volunteers. Canarypox is the first candidate HIV vaccine that has induced cross-clade functional CTL responses. The first phase I trial of the candidate vaccine in Africa was launched early in 1999 with Ugandan volunteers. The study determined the extent to which Ugandan volunteers have CTL that are active against the subtypes of HIV prevalent in Uganda, A and D.

Other strategies that have progressed to phase I trials in uninfected persons include peptides, lipopeptides, DNA, an attenuated Salmonella vector, lipopeptides, p24, etc. Specifically, candidate vaccines that induce one or more of the following are being sought:

- neutralizing antibodies active against a broad range of HIV primary isolates;
- cytotoxic T cell responses in a vast majority of recipients;
- strong mucosal immune responses.

Phase II

On December 13, 2004, the HIV Vaccine Trials Network (HVTN) began recruiting for the STEP study, a 3,000-participant phase II clinical trial of a novel HIV vaccine, at sites in North America, South America, the Caribbean and Australia. The trial was co-funded by the National Institute of Allergy and Infectious Diseases (NIAID), which is a division of the National Institutes of Health (NIH), and the pharmaceutical company Merck & Co. Merck developed the experimental vaccine called V520 to stimulate HIV-specific cellular immunity, which prompts the body to produce T cells that kill HIV-infected cells. In previous smaller trials, this vaccine was found to be safe, because of the lack of adverse effects on the patients. The vaccine showed induced cellular immune responses against HIV in more than half of volunteers.

V520 contains a weakened adenovirus that serves as a carrier for three subtype B HIV genes. Subtype B is the most prevalent HIV subtype in the regions of the study sites. Adenoviruses are among the main causes of upper respiratory tract ailments such as the common cold. Because the vaccine contains only three HIV genes housed in a weakened adenovirus, study participants cannot become infected with HIV or get a respiratory infection from the vaccine.

It was announced in September 2007 that the trial for V520 would be discontinued after it determined that the vaccination was ineffective. The foremost issue facing the rAd5 adenovirus that was used is the high prevalence of the adenovirus-specific antibodies as a result of prior exposure to the virus. Adenovirus vectors and many other viral vectors currently used in HIV vaccines, will induce a rapid memory immune response against the vector. This results in an impediment to the development of a T cell response against the inserted antigen (HIV antigens) Additionally, it appears that V520 may have made some recipients more receptive to infection by HIV-1.

The HVTN expected to finish the study in 2009, but ceased further treatment administration and declared the vaccine ineffective at preventing HIV-infection in September 2007. The results of the trial have caused some to call for a reexamination of vaccine development strategies.

Phase III

In February 2003, VaxGen announced that their AIDSVAX vaccine was a failure in North America as there was not a statistically significant reduction of HIV infection within the study population. This same vaccine was retested in Thailand within a vaccine regimen called RV 144 beginning in 2003, with positive results. In both cases the vaccines targeted gp120 and were specific for the geographical regions. The Thai trial was the largest AIDS vaccine trial to date when it started.

In October 2009, the results of the RV 144 trial were published. Initial results, released in September 2009 prior to publication of complete results, were encouraging for scientists in search of a vaccine. The study involved 16,395 participants who did not have HIV infection, 8197 of whom were given treatment consisting of two experimental vaccines targeting HIV types B and E that are prevalent in Thailand, while 8198 were given a placebo.

The participants were tested for HIV every six months for three years. After three years, the vaccine group saw HIV infection rates reduced by more than 30% compared with those in the placebo group. However, after taking into account the seven people who had HIV infections at the time of their vaccination (two in the placebo group, five in the vaccine group) the percentage dropped to 26%.

Planned Clinical Trials

Novel approaches, including modified vaccinia Ankara (MVA), adeno-associated virus, Venezuelan equine encephalitis (VEE) replicons, and codon-optimized DNA have proven to be strong inducers of CTL in macaque models, and have provided at least partial protection in some models. Most of these approaches are in, or will soon enter, clinical studies.

Economics of Vaccine Development

A June 2005 study estimates that $682 million is spent on AIDS vaccine research annually. Economic issues with developing an AIDS vaccine include the need for advance purchase commitment (or advance market commitments) because after an AIDS vaccine has been developed, governments and NGOs may be able to bid the price down to marginal cost.

Classification of all Theoretically Possible HIV Vaccines

Any theoretically possible HIV vaccines must inhibit or stop the HIV virion replication cycle.So, the targets of the vaccine are the following phases of the HIV virion cycle:

- Phase I. Free state
- Phase II. Attachment
- Phase III. Penetration
- Phase IV. Uncoating
- Phase V. Replication
- Phase VI. Assembling
- Phase VII. Releasing.

So, the possible approaches for the HIV vaccine are the following (in the bracket specified the *Phases* were it is possible to do).

Filtering of the Virions from Blood (Phase I)

- Biological approach for removing the HIV virions from the blood.
- Chemical approach for removing the HIV virions from the blood.
- Physical approach for removing the HIV virions from the blood.

Different Approaches to Catch the Virion (Phase I-III, VI, VII)

- Phagocytosis of the HIV virions.
- Chemical or organic based capture (creation of any skin or additional membrane around the virion) of HIV virions
- Chemical or organic attachments to the virion.

Different Approaches to Destroy or Damage the Virion or its Parts (Phase I-VII)

Here, "damage" means inhibiting or stopping the ability of virion to process any of the *Phase II-VII*. Here are the different classification of methods:

- By nature of method:
  - o Physical methods (*Phase I-VII*)
  - o Chemical and biological methods (*Phase I-VII*)
- By damaging target of the HIV virion structure:
  - o Damaging the Docking Glycoprotein gp120 (*Phase I-III, VI, VII*)
  - o Damaging the Transmembrane Glycoprotein gp41 (*Phase I-III, VI, VII*)
  - o Damaging the virion matrix (*Phase I-III, VI, VII*)
  - o Damaging the virion Capsid (*Phase I-III, VI, VII*)
  - o Damaging the Reverse Transcriptase (*Phase I-VII*)
  - o Damaging the RNA (*Phase I-VII*).

Blocking the Replication (Phase I)

- Insertion into blood chemical or organic compounds which binds to the gp120. Hypothetically, it can be pieces of the CD4 cell membranes with receptors. Any chemical and organic alternative (with ability to bind the gp120) of this receptors also can be used.

- Insertion into blood chemical or organic compounds which binds to the receptors of the CD4 cells.

Inhibiting Process of Phases (Drugs Already used for this Approach)

- Biological, chemical or physical approach to inhibit the *Attachment*
- Biological, chemical or physical approach to inhibit the *Penetration*
- Biological, chemical or physical approach to inhibit the *Uncoating* including introducing the mutation into the HIV
- Biological, chemical or physical approach to inhibit the *Replication* including introducing the mutation into the HIV
- Biological, chemical or physical approach to inhibit the *Assembling* including introducing the mutation into the HIV
- Biological, chemical or physical approach to inhibit (capping) the *Releasing*.

Methods of the Inhibiting of the Functionality of the Infected Cell (Phase VI-VII)

Inhibiting the life functions of the infected cell:

- Inhibiting the metabolism of the infected cell
- Inhibiting the energy exchange of the infected cell.

Future Work

According to Gary J. Nabel of the Vaccine Research Centre in Bethesda, Maryland, several hurdles must be overcome before scientific research will culminate in a definitive AIDS vaccine. First, greater translation between animal models and human trials must be established. Second, new, more effective, and more easily produced vectors must be identified. Finally, and most importantly, there must arise a robust understanding of the immune response to potential vaccine candidates. Emerging technologies that enable the identification of T-cell-receptor specificities and cytokine profiles will prove invaluable in hastening this process.

5

Biological Warfare : A Serious Threat in the 21st Century

Biological warfare (BW), also known as germ warfare, is the deliberate use of disease-causing biological agents such as protozoa, fungi, bacteria, protists, or viruses, to kill or incapacitate humans, other animals or plants. Biological weapons (often referred to as *bioweapons*) are living organisms or replicating entities (virus) that reproduce or replicate within their host victims. Biological weapons may be employed in various ways to gain a strategic or tactical advantage over an adversary, either by threat or by actual deployment. Like some of the chemical weapons, biological weapons may also be useful as area denial weapons. These agents may be lethal or non-lethal, and may be targeted against a single individual, a group of people, or even an entire population. They may be developed, acquired, stockpiled or deployed by nation states or by non-national groups. In the latter case, or if a nation-state uses it clandestinely, it may also be considered bioterrorism.

There is an overlap between biological warfare and chemical warfare, as the use of toxins produced by living organisms is considered under the provisions of both the Biological Weapons Convention and the Chemical Weapons Convention. Toxins and Psychochemical weapons are often referred to as *midspectrum agents*. Unlike bioweapons, these midspectrum agents do not reproduce in their host and are typically characterized by shorter incubation periods.

Offensive biological warfare, including mass production, stockpiling and use of biological weapons, was outlawed by the 1972 Biological Weapons Convention (BWC). The rationale behind this treaty, which has been ratified or acceded to by 163 countries as of 2009, is to prevent a biological attack which could conceivably result in large numbers of civilian fatalities and cause severe disruption to economic and societal infrastructure. Many countries, including signatories of the BWC, currently pursue research into the defence or protection against BW, which is not prohibited by the BWC.

A nation or group that can pose a credible threat of mass casualty has the ability to alter the terms on which other nations or groups interact with it. Biological weapons allow for the potential to create a level of destruction and loss of life far in excess of nuclear, chemical or conventional weapons, relative to their mass and cost of development and storage. Therefore, biological agents may be useful as strategic deterrents in addition to their utility as offensive weapons on the battlefield. As a tactical weapon for military use, a significant problem with a BW attack is that it would take days to be effective, and therefore might not immediately stop an opposing force. Some biological agents (especially smallpox, plague, and tularemia) have the capability of person-to-person transmission via aerosolized respiratory droplets. This feature can be undesirable, as the agent(s) may be transmitted by this mechanism to unintended populations, including neutral or even friendly forces. While containment of BW transmission is less of a concern for certain criminal or terrorist organizations, it remains a significant concern for the military and civilian populations of virtually all nations.

Characteristics of Biological Weapons

Anti-personnel

Ideal characteristics of a biological agent to be used as a weapon against humans are high infectivity, high virulence, non-availability of vaccines, and availability of an effective and efficient delivery system. Stability of the weaponized agent (ability of the agent to retain its infectivity and virulence after a prolonged period of storage) may also be desirable, particularly for military applications.

The primary difficulty is not the production of the biological agent, as many biological agents used in weapons can often be manufactured relatively quickly, cheaply and easily. Rather, it is the weaponization, storage and delivery in an effective vehicle to a vulnerable target that pose significant problems.

For example, *Bacillus anthracis* is considered an effective agent for several reasons. First, it forms hardy spores, perfect for dispersal aerosols. Second, this organism is not considered transmissible from person to person, and thus rarely if ever causes secondary infections. A pulmonary anthrax infection starts with ordinary influenza-like symptoms and progresses to a lethal hemorrhagic mediastinitis within 3–7 days, with a fatality rate that is 90% or higher in untreated patients. Finally, friendly personnel can be protected with suitable antibiotics.

A large-scale attack using anthrax would require the creation of aerosol particles of 1.5 to 5 microns. Too large and the particles would not reach the lower respiratory tract. Too small and the particles would be exhaled back out into the atmosphere. At this size, conductive powders tend to aggregate because of electrostatic charges, hindering dispersion. So the material must be treated to insulate and neutralize the charges. The weaponized agent must be resistant to degradation by rain and ultraviolet radiation from sunlight, while retaining the ability to efficiently infect the human lung. There are other technological difficulties as well, chiefly relating to storage of the weaponized agent.

Agents considered for weaponization, or known to be weaponized, include bacteria such as *Bacillus anthracis*, *Brucella spp.*, *Burkholderia mallei*, *Burkholderia pseudomallei*, *Chlamydophila psittaci*, *Coxiella burnetii*, *Francisella tularensis*, some of the Rickettsiaceae (especially *Rickettsia prowazekii* and *Rickettsia rickettsii*), *Shigella spp.*, *Vibrio cholerae*, and *Yersinia pestis*. Many viral agents have been studied and/or weaponized, including some of the Bunyaviridae (especially Rift Valley fever virus), Ebolavirus, many of the Flaviviridae (especially Japanese encephalitis virus), Machupo virus, Marburg virus, Variola virus, and Yellow fever virus. Fungal agents that have been studied include *Coccidioides spp.*.

Toxins that can be used as weapons include ricin, staphylococcal enterotoxin B, botulinum toxin, saxitoxin, and many mycotoxins. These toxins and the organisms that produce them are sometimes referred to as select agents. In the United States, their possession, use, and transfer are regulated by the Centres for Disease Control and Prevention's Select Agent Program.

Anti-agriculture

Biological warfare can also specifically target plants to destroy crops or defoliate vegetation. The United States and Britain discovered plant growth regulators (i.e., herbicides) during the Second World

War, and initiated an herbicidal warfare program that was eventually used in Malaya and Vietnam in counter insurgency. Though herbicides are chemicals, they are often grouped with biological warfare as bioregulators in a similar manner as biotoxins. Scorched earth tactics or destroying livestock and farmland were carried out in the Vietnam war (cf. Agent Orange) and Eelam War in Sri Lanka.

The United States developed an anti-crop capability during the Cold War that used plant diseases (bioherbicides, or mycoherbicides) for destroying enemy agriculture. It was believed that destruction of enemy agriculture on a strategic scale could thwart Sino-Soviet aggression in a general war. Diseases such as wheat blast and rice blast were weaponized in aerial spray tanks and cluster bombs for delivery to enemy watersheds in agricultural regions to initiate epiphytotics (epidemics among plants). When the United States renounced its offensive biological warfare program in 1969 and 1970, the vast majority of its biological arsenal was composed of these plant diseases.

In 1980s Soviet Ministry of Agriculture had successfully developed variants of foot-and-mouth disease, and rinderpest against cows, African swine fever for pigs, and psittacosis to kill chicken. These agents were prepared to spray them down from tanks attached to airplanes over hundreds of miles. The secret program was code-named "Ecology".

Attacking animals is another area of biological warfare intended to eliminate animal resources for transportation and food. In the First World War, German agents were arrested attempting to inoculate draft animals with anthrax, and they were believed to be responsible for outbreaks of glanders in horses and mules. The British tainted small feed cakes with anthrax in the Second World War as a potential means of attacking German cattle for food denial, but never employed the weapon. In the 1950s, the United States had a field trial with hog cholera. During the Mau Mau Uprising in 1952, the poisonous latex of the African milk bush was used to kill cattle.

Unconnected with inter-human wars, humans have deliberately introduced the rabbit disease Myxomatosis, originating in South America, to Australia and Europe, with the intention of reducing the rabbit population- which had devastating but temporary results, with wild rabbit populations reduced to a fraction of their former size but survivors developing immunity and increasing again.

Biodefence

Biodefence refers to short term, local, usually military measures to restore biosecurity to a given group of persons in a given area — in the civilian terminology, it is a very robust biohazard response. It is technically possible to apply biodefence measures to protect animals or plants, but this is generally uneconomic. However, protection of water supplies and food supplies are often a critical part of biodefence. Various definitions of biosafety emerged in different professions to guarantee non-human health. Biodefence is most often discussed in the context of biowar or bioterrorism, and is generally considered a military or emergency response term.

Biodefence applies to two distinct target populations: civilian non-combatant and military combatant (troops in the field).

Biodefence of Troops in the Field

The Department of Defence (or "DoD") has focused since at least 1998 on the development and application of vaccine-based biodefences. In a July 2001 report commissioned by the DoD, the "DoD-critical products" were stated as vaccines against anthrax (AVA and Next Generation), smallpox, plague, tularemia, botulinum, ricin, and equine encephalitis. Note that two of these targets are toxins (botulinum and ricin) while the remainder are infectious agents.

Role of Public Health Departments and Disease Surveillance

It is important to note that all of the classical and modern biological weapons organisms are animal diseases, the only exception being smallpox. Thus, in any use of biological weapons, it is highly likely that animals will become ill either simultaneously with, or perhaps earlier than humans. Indeed, in the largest biological weapons accident known– the anthrax outbreak in Sverdlovsk (now Yekaterinburg) in the Soviet Union in 1979, sheep became ill with anthrax as far as 200 kilometres from the release point of the organism from a military facility in the southeastern portion of the city. Thus, a robust surveillance system involving human clinicians and veterinarians may identify a bioweapons attack early in the course of an epidemic, permitting the prophylaxis of disease in the vast majority of people (and/or animals) exposed but not yet ill.

For example in the case of anthrax, it is likely that by 24 – 36 hours after an attack, some small percentage of individuals (those with compromised immune system or who had received a large dose of the organism due to proximity to the release point) will become ill with classical symptoms and signs (including a virtually unique chest X-ray finding, often recognized by public health officials if they receive timely reports). By making these data available to local public health officials in real time, most models of anthrax epidemics indicate that more than 80% of an exposed population can receive antibiotic treatment before becoming symptomatic, and thus avoid the moderately high mortality of the disease.

Identification of Bioweapons

The goal of biodefence is to integrate the sustained efforts of the national and homeland security, medical, public health, intelligence, diplomatic, and law enforcement communities. Health care providers and public health officers are among the first lines of defence. In some countries private, local, and provincial (state) capabilities are being augmented by and coordinated with federal assets, to provide layered defences against biological weapons attacks. During the first Gulf War the United Nations activated a biological and chemical response team, Task Force Scorpio, to respond to any potential use of weapons of mass destruction on civilians.

The traditional approach toward protecting agriculture, food, and water: focusing on the natural or unintentional introduction of a disease is being strengthened by focused efforts to address current and anticipated future biological weapons threats that may be deliberate, multiple, and repetitive. The growing threat of biowarfare agents and bioterrorism has led to the development of specific field tools that perform on-the-spot analysis and identification of encountered suspect materials. One such technology, being developed by researchers from the Lawrence Livermore National Laboratory (LLNL), employs a "sandwich immunoassay", in which fluorescent dye-labelled antibodies aimed at specific pathogens are attached to silver and gold nanowires.

In the Netherlands, the company TNO has designed Bioaerosol Single Particle Recognition eQuipment (BiosparQ). This system would be implemented into the national response plan for bioweapons attacks in the Netherlands.

Researchers at Ben Gurion University in Israel are developing a different device called the BioPen, essentially a "Lab-in-a-Pen", which can detect known biological agents in under 20 minutes using an adaptation of the ELISA, a similar widely employed immunological technique, that in this case incorporates fibre optics.

History

Biological warfare has been practiced repeatedly throughout history. Before the 20th century, the use of biological agents took three major forms:

- Deliberate poisoning of food and water with infectious material
- Use of microorganisms, toxins or animals, living or dead, in a weapon system
- Use of biologically inoculated fabrics.

Antiquity

The earliest documented incident of the intention to use biological weapons is recorded in Hittite texts of 1500–1200 B.C, in which victims of plague were driven into enemy lands. Although the Assyrians knew of ergot, a parasitic fungus of rye which produces ergotism when ingested, there is no evidence that they poisoned enemy wells with the fungus, as has been claimed.

According to Homer's epic poems about the legendary Trojan War, the *Iliad* and the *Odyssey,* spears and arrows were tipped with poison. During the First Sacred War in Greece, in about 590 BC, Athens and the Amphictionic League poisoned the water supply of the besieged town of Kirrha (near Delphi) with the toxic plant hellebore. The Roman commander Manius Aquillus poisoned the wells of besieged enemy cities in about 130 BC.

During the 4th century BC Scythian archers tipped their arrow tips with snake venom, human blood, and animal feces to cause wounds to become infected. There are numerous other instances of the use of plant toxins, venoms, and other poisonous substances to create biological weapons in antiquity.

In 184 B.C, Hannibal of Carthage had clay pots filled with venomous snakes and instructed his soldiers to throw the pots onto the decks of Pergamene ships. In about AD 198, the Parthian city of Hatra (near Mosul, Iraq) repulsed the Roman army led by Septimius Severus by hurling clay pots filled with live scorpions at them.

Middle Ages

The Mongol Empire established commercial and political connections between the Eastern and Western areas of the world, but it did so through the most mobile army ever seen. The armies, being the most rapidly moving travelers who had ever moved between the steppes of East Asia (where bubonic plague was and remains endemic among small rodents) and managed to keep the chain of infection without a break until they reached, and infected, peoples and rodents who had never encountered it. The ensuing Black Death may have killed almost half of the population of Europe in the next decades, changing the course of Asian and European history.

During the Middle Ages, victims of the bubonic plague were used for biological attacks, often by flinging the infected corpses (formites)and excrement over castle walls using catapults. In 1346, during the siege of Kafa (now Feodossia Ukraine) the attacking Tartar Forces which were subjugated by the Mongol empire under Genghis Khan, used the bodies of Mongol warriors of the Golden Horde who had died of plague, as weapons. The dead warriors were thrown over the walls of the besieged city of Crimean. An outbreak of plague followed and the defending forces retreated, followed by the conquest of the city by the Mongol army. It has been speculated that this operation may have been responsible for the advent of the Black Death in Europe.

At the siege of Thun-l'Eveque in 1340, during the Hundred Years' War, the attackers catapulted decomposing animals into the besieged area.

In 1422, during the siege of Karlstein Castle in Bohemia, Hussite attackers used catapults to throw dead (but not plague-infected) bodies and 2000 carriage-loads of dung over the walls.

The last known incident of using plague corpses for biological warfare occurred in 1710, when Russian forces attacked the Swedes by flinging plague-infected corpses over the city walls of Reval (Tallinn). However, during the 1785 siege of La Calle, Tunisian forces flung diseased clothing into the city.

18th Century

The Native American population was devastated after contact with the Old World due to the introduction of many different fatal diseases. There are two documented cases of alleged and attempted germ warfare. The first, during a parley at Fort Pitt on June 24, 1763, Ecuyer gave representatives of the besieging Delawares two blankets and a handkerchief that had been exposed to smallpox, hoping to spread the disease to the Natives in order to end the siege. William Trent, the militia commander, left records that clearly indicated that the purpose of giving the blankets was "to Convey the Smallpox to the Indians."

British commander Lord Jeffrey Amherst and Swiss-British officer Colonel Henry Bouquet certainly discussed this, in the course of Pontiac's Rebellion; there still exists correspondence referencing the idea of giving smallpox-infected blankets to enemy Indians. Historian Francis Parkman verifies four letters from June 29, July 13, 16 and 26th, 1763. Excerpts: Commander Lord Jeffrey Amherst writes July 16, 1763, "P.S. You will Do well to try to Inocculate the Indians by means of Blankets, as well as to try Every other method that can serve to Extirpate this Execrable Race. I should be very glad your Scheme for Hunting them Down by Dogs could take Effect,..." Colonel Henry Bouquet replies July 26, 1763, "I received yesterday your Excellency's letters of 16th with their Inclosures. The signal for Indian Messengers, and all your directions will be observed."

While the intent of carrying out biological warfare is clear, there is debate among historians as to whether this actually took place despite Bouquet's affirmative reply to Amherst, and the continuing correspondence on the point. Smallpox is highly infectious and does not require contaminated blankets to spread uncontrollably, and together with measles, influenza, chicken pox, etc. had been doing so since the arrival of Europeans and their animals. Historians have been unable to establish whether or not the Amherst plan was implemented, particularly in light of the fact that smallpox was already present in the region, and that scientific knowledge of disease at that time had yet to develop an understanding of infection vectors, nor in the case of smallpox a full acknowledgment of the protective effect of a cowpox infection.

Regardless of whether the plan was carried out, trade and combat provided ample opportunity for transmission of the disease.

The diseases which depopulated the New World can be traced to Eurasia where people had long lived with them and developed some immunological ability to survive their presence. Without similarly long ancestral exposure, indigenous Americans were immunologically naive and extremely vulnerable.

19th Century

In 1834, Massachusetts diarist Richard Henry Dana visited San Francisco on a merchant ship. His ship traded many items including blankets with Mexicans and Russians who had established outposts on the northern side of the San Francisco Bay. Local histories document that the California plague epidemic began at the Russian fort soon after they left. It is possible that the blankets were the source of the contamination (hidden fleas, or rats, perhaps), but another possible source was a Chinese ship making port in San Francisco at the same time. Plague became established in California and has since become endemic throughout much of the North American West. Native rodents have suffered a severe population decline, only partly due to human eradication action. During the American Civil War, General Sherman reported that Confederate forces shot farm animals in ponds upon which the Union troops depended for drinking water. This would have made the water unpleasant to drink, though perhaps the death caused might not have been that desired. A Confederate doctor planned and may have carried out a bacteriological attack on Northern populations across the Canadian border.

Jack London, in his story "'Yah! Yah! Yah!'", described a punitive European expedition to a South Pacific island deliberately exposing the Polynesian population to measles, of which many of them died. While much of the material for London's *South Sea Tales* is derived from his personal experience in the region, it is not known whether this particular incident is historical.

20th Century

During the First World War, the Empire of Germany pursued an ambitious biological warfare program. Using diplomatic pouches and couriers, the German General Staff supplied small teams of saboteurs in the Russian Duchy of Finland, and in the then-neutral countries of Romania, the United States, and Argentina.

In Finland, saboteurs mounted on reindeer placed ampoules of anthrax in stables of Russian horses in 1916. Anthrax was also supplied to the German military attache in Bucharest, as was glanders, which was employed against livestock destined for Allied service.

German intelligence officer and US citizen Dr. Anton Casimir Dilger established a secret lab in the basement of his sister's home in Chevy Chase, Maryland, that produced glanders which was used to infect livestock in ports and inland collection points including, at least, Newport News, Norfolk, Baltimore, and New York, and probably St. Louis and Covington, Kentucky. In Argentina, German agents also employed glanders in the port of Buenos Aires and also tried to ruin wheat harvests with a destructive fungus.

The Geneva Protocol of 1925 prohibited the use of chemical weapons and biological weapons, but said nothing about experimentation, production, storage, or transfer; later treaties did cover these aspects. Twentieth-century advances in microbiology enabled the first pure-culture biological agents to be developed by World War II.

The interwar period was a period of development by many nations, most notably the Empire of Japan. Secret Imperial Japanese Army Unit 731, based primarily at Pingfan in Manchuria commanded by Lieutenant General Shirô Ishii, did research on BW, conducted often fatal human experiments on prisoners, and produced biological weapons for combat use during the Second Sino-Japanese War.

Biological experiments, often using twins with one subject to the procedure and the other as a control, were carried out by Nazi Germany on concentration camp inmates, particularly by Joseph Mengele.

1937–1945

During the Sino-Japanese War (1937–1945) and World War II, the Imperial Japanese Army made use of biological weapons against both Chinese soldiers and civilians in several military campaigns. Three veterans of Unit 731 testified, in a 1989 interview to the Asahi Shimbun, that they were part of a mission to contaminate the Horustein river with typhoid near the Soviet troops during the Battle of Khalkhin Gol. In 1940, the Imperial Japanese Army Air Force bombed Ningbo with ceramic bombs full of fleas carrying the bubonic plague. A film showing this operation was seen by the imperial princes Tsuneyoshi Takeda and Takahito Mikasa during a screening made by mastermind Shiro Ishii. During the Khabarovsk War Crime Trials the accused, such as Major General Kiyashi Kawashima, testified that as early as 1941 some 40 members of Unit 731 air-dropped plague-contaminated fleas on Changde. These operations caused epidemic plague outbreaks.

Many operations were ineffective due to inefficient delivery systems, using disease-bearing insects rather than dispersing the agent as an bioaerosol cloud. Nevertheless, some modern Chinese historians estimate that 400,000 Chinese died as a direct result of Japanese field testing and operational use of biological weapons.

In response to biological weapons development in Japan, and at the time suspected in Nazi Germany, the United States, United Kingdom, and Canada initiated a BW development programs in 1941 that resulted in the weaponization of tularemia, anthrax, brucellosis, and botulism toxin. The centre for United States military BW research was Fort Detrick, Maryland, where USAMRIID is currently based;

the first director was pharmaceutical executive George W. Merck. Some biological and chemical weapons research and testing was also conducted at Dugway Proving Grounds in Utah, at a munition manufacturing complex in Terre Haute, Indiana, and at a tract on Horn Island, Mississippi.

Much of the British work was carried out at Porton Down. Field testing carried out in the United Kingdom during World War II left Gruinard island in Scotland contaminated with anthrax for the next 48 years.

1946 to 1972

During the 1948 Israel War of Independence, International Red Cross reports raised suspicion that the Jewish Haganah militia had released Salmonella typhi bacteria into the water supply for the city of Acre, causing an outbreak of typhoid among the inhabitants. Egyptian troops later claimed to have captured disguised Haganah soldiers near wells in Gaza, whom they executed for allegedly attempting another attack. Israel denies these allegations.

During the Cold War, US conscientious objectors were used as consenting test subjects for biological agents in a program known as Operation Whitecoat. There were also many unpublicized tests carried out on the public during the Cold War.

Considerable research on the topic was performed by the United States, the Soviet Union, and probably other major nations throughout the Cold War era, though it is generally believed that biological weapons were never used after World War II. This view was challenged by China and North Korea, who accused the United States of germ warfare in the Korean War (1950–1953).

Cuba has also accused the United States of spreading human and animal disease on their island nation.

At the time of the Korean War the United States had only weaponized one agent, brucellosis ("Agent US"), which is caused by *Brucella suis*. The original weaponized form used the M114 bursting bomblet in M33 cluster bombs.

While the specific form of the biological bomb was classified until some years after the Korean War, in the various exhibits of biological weapons that Korea alleged were dropped on their country nothing resembled an M114 bomblet. There were ceramic containers that had some similarity to Japanese weapons used against the Chinese in

World War II, developed by Unit 731. Some of the Unit 731 personnel were imprisoned by the Soviets, and would have been a potential source of information on Japanese weaponization. The head of Unit 731, Lieutenant General Shiro Ishii, was granted immunity from war crimes prosecution in exchange for providing information to the United States on the Unit's activities.

The Korean War allegations also stressed the use of disease vectors, such as fleas, which, again, were probably a legacy of Japanese biological warfare efforts. The United States initiated its weaponization efforts with disease vectors in 1953, focused on Plague-fleas, EEE-mosquitoes, and yellow fever- mosquitoes (OJ-AP). However, US medical scientists in occupied Japan undertook extensive research on insect vectors, with the assistance of former Unit 731 staff, as early as 1946.

The United States Air Force was not satisfied with the operational qualities of the M114/US and labelled it an interim item until the United States Army Chemical Corps could deliver a superior weapon. The Air Force also changed its plans and wanted lethal biologicals.

The Chemical Corps then initiated a crash program to weaponize anthrax (N) in the E61 1/2-lb hour-glass bomblet. Though the program was successful in meeting its development goals, the lack of validation on the infectivity of anthrax stalled standardization.

Around 1950 the Chemical Corps also initiated a program to weaponize tularemia (UL). Shortly after the E61/N failed to make standardization, tularemia was standardized in the 3.4" M143 bursting spherical bomblet. This was intended for delivery by the MGM-29 Sergeant missile warhead and could produce 50% infection over a 7-square-mile (18 km^2) area.

Unlike anthrax, tularemia had a demonstrated infectivity with human volunteers (Operation Whitecoat). Furthermore, although tularemia is treatable by antibiotics, treatment does not shorten the course of the disease.

In addition to the use of bursting bomblets for creating biological aerosols, the Chemical Corps started investigating aerosol-generating bomblets in the 1950s. The E99 was the first workable design, but was too complex to be manufactured. By the late 1950s the 4.5" E120 spraying spherical bomblet was developed; a B-47 bomber with a SUU-24/A dispenser could infect 50% or more of the population of a 16-square-mile (41 km^2) area with tularemia with the E120. The E120 was later superseded by dry-type agents. Dry-type biologicals resemble talcum powder, and can be disseminated as aerosols using gas expulsion devices instead of a burster or complex sprayer. The Chemical Corps developed Flettner rotor bomblets and later triangular bomblets for wider coverage due to improved glide angles over Magnus-lift spherical bomblets. Weapons of this type were in advanced development by the time the program ended.

United States President Richard Nixon signed an executive order on November 1969, which stopped production of biological weapons in the United States and allowed only scientific research of lethal biological agents and defensive measures such as immunization and biosafety. The biological munition stockpiles were destroyed, and approximately 2,200 researchers became redundant.

United States Special Forces and the CIA also had an interest in biological warfare, and a series of special munitions was created for their operations. The covert weapons developed for the military (M1, M2, M4, M5, and M32- or Big Five Weapons) were destroyed in accordance with Nixon's executive order to end the offensive program. The CIA maintained its collection of biologicals well into 1975 when it became the subject of the senate Church Committee.

The Biological Weapons Convention

In 1972, the United States signed the Biological and Toxic Weapons Convention, which banned the "development, production and stockpiling of microbes or their poisonous products except in amounts necessary for protective and peaceful research." By 1996, 137 countries had signed the treaty; however it is believed that since the signing of the Convention the number of countries capable of producing such weapons has increased.

The Soviet Union continued research and production of offensive biological weapons in a program called Biopreparat, despite having signed the convention. The United States was unaware of the program until Dr. Vladimir Pasechnik defected in 1989, and Dr. Kanatjan Alibekov, the first deputy director of Biopreparat defected in 1992.

After the 1991 Persian Gulf War, Iraq admitted to the United Nations inspection team to having produced 19,000 litres of concentrated botulinum toxin, of which approximately 10,000 L were loaded into military weapons; the 19,000 litres have never been fully accounted for. This is approximately three times the amount needed to kill the entire current human population by inhalation, although in practice it would be impossible to distribute it so efficiently, and, unless it is protected from oxygen, it deteriorates in storage.

On September 18, 2001 and for a few days after several letters were received by members of the United States Congress and media outlets containing anthrax spores: the attack killed five people. The identity of the perpetrator remained unknown until 2008, when a primary suspect was named.

List of BW Institutions and Programs by Country

According to the United States Office of Technology Assessment, since disbanded, seventeen countries were believed to possess biological weapons in 1995: Libya, North Korea, South Korea, Iraq, Taiwan, Syria, Israel, Iran, China, Egypt, Vietnam, Laos, Cuba, Bulgaria, India, South Africa, and Russia.

United States

The United States biological weapons program officially began in the spring 1943 on orders from U.S. President Franklin Roosevelt. Research continued following World War II as the U.S. built up a large stockpile of biological agents and weapons. Throughout its history the program was secret. It became controversial when it was later revealed that laboratory and field testing (some of the latter using simulants on non-consenting individuals) had been common. The official policy of the United States was first to deter the use of bio-weapons against U.S. forces and secondarily to retaliate if deterrence failed. There exists no evidence that the U.S. ever used biological agents against an enemy in the field.

In 1969, President Richard Nixon ended all offensive (i.e., non-defensive) aspects of the U.S. bio-weapons program. In 1975 the U.S. ratified both the 1925 Geneva Protocol and the 1972 Biological Weapons Convention (BWC) — these are international treaties outlawing biological warfare. Recent U.S. biodefence programs, however, have raised concerns that the US may be pursuing research that is outlawed by the BWC.

History

Early History (1918-41)

The United States' first interest in any form of biological warfare came at the close of World War I. The only agent the U.S. tested was the toxin ricin. The U.S. conducted tests concerning two methods of ricin dissemination, the first, involved adhering the toxin to shrapnel for delivery by artillery shell, which was successful. The other method, delivering an aerosol cloud of ricin, proved less successful. Neither delivery method was perfected before the war in Europe ended.

In the early 1920s suggestions that the U.S. begin a biological weapons program were coming from within the Chemical Warfare Service (CWS). Chief of the CWS, Amos Fries, decided that such a program would not be "profitable" for the U.S. Japan's Shiro Ishii began promoting biological weapons during the 1920s and toured biological research facilities worldwide, including in the United States. Though Ishii concluded that the U.S. was developing a bio-weapons programs he was incorrect. In fact, Ishii concluded that all major powers he visited was developing a bio-weapons program. As the interwar period continued, the United States did not emphasize biological weapons development or research. While the U.S. was spending very little time on biological weapons research, its future allies and enemies in the upcoming second World War were researching the potential of biological weapons as early as 1933.

World War II (1941-45)

Despite the World War I-era interest in ricin, as World War II erupted the United States Army still maintained the position that biological weapons was, for the most part, impractical. Other nations, notably France, Japan and the United Kingdom, thought otherwise and had begun their own biological weapons programs. Thus, as late as 1942 the U.S. had no biological weapons capabilities. Initial interest in biological weapons by the Chemical Warfare Service began in 1941. That fall, U.S. Secretary of War Henry L. Stimson requested that the National Academy of Sciences (NAS) undertake consideration of U.S. biological warfare. He wrote to Dr. Frank B. Jewett, then president of the NAS:

Because of the dangers that might confront this country from potential enemies employing what may be broadly described as biological warfare, it seems advisable that investigations be initiated to survey the present situation and the future possibilities. I am therefore, asking if you will undertake the appointment of an appropriate committee to survey all phases of this matter. Your organization already has before it a request from The Surgeon General for the appointment of a committee by the Division of Medical Sciences of the National Research Council to examine one phase of the matter.

In response the NAS formed a committee, the War Bureau of Consultants (WBC), which issued a report on the subject in February 1942. The report, among other items, recommended the research and development of an offensive biological weapons program.

The British, and the research undertaken by the WBC, pressured the U.S. to begin biological weapons research and development and in November 1942 U.S. President Franklin Roosevelt officially approved an American biological weapons program. In response to the information provided by the WBC, Roosevelt ordered Stimson to form the War Research Service (WRS). Established within the Federal Security Agency, the WRS' stated purpose was to promote "public security and health", but, in reality, the WRS was tasked with coordinating and supervising the U.S. biological warfare program. In the spring of 1943 the U.S. Army Biological Warfare Laboratories were established at Fort (then Camp) Detrick in Maryland.

Though initially, under George Merck, the WRS contracted several universities to participate in the U.S. biological weapons program, the program became large quickly and before long it was under the full control of the CWS. By November 1943 the biological weapons facility at Detrick was completed, in addition, the United States constructed three other facilities- a biological agent production plant at Vigo County near Terre Haute, Indiana, a field-testing site on Horn Island in Mississippi, and another field site near Granite Peak in Utah. According to an official history of the period, "the elaborate security precautions taken [at Camp Detrick] were so effective that it was not until January 1946, 4 months after VJ Day, that the public learned of the wartime research in biological weapons".

Cold War (1946-69)

Immediately following World War II, production of U.S. biological warfare agents went from "factory-level to laboratory-level". Meanwhile, work on biological weapons delivery systems increased. By 1950 the principal U.S. bio-weapons facility was located at Camp Detrick in Maryland under the auspices of the Research and Engineering Division of the U.S. Army Chemical Corps. The U.S. also maintained bio-warfare facilities at Fort Terry, an animal research facility on Plum Island. From the end of World War II through the Korean War, the U.S. Army, the Chemical Corps and the U.S. Air Force all made great strides in their biological warfare programs, especially concerning delivery systems.

The U.S. biological program expanded significantly during the Korean War. From 1952-1954 the Chemical Corps maintained a biological weapons research and development facility at Fort Terry on Plum Island, New York. The Fort Terry facility's focus was on anti-animal biological weapon research and development; the facility researched more than a dozen potential BW agents. A facility was opened in Pine Bluff, Arkansas, Pine Bluff Arsenal and by 1954 the production of weapons-grade agents began.

End of the Program (1969-73)

President Richard M. Nixon issued his "Statement on Chemical and Biological Defence Policies and Programs" on November 25, 1969 in a speech from Fort Detrick. The statement ended, unconditionally, all U.S. offensive biological weapons programs. Nixon noted that biological weapons were unreliable and stated:

The United States shall renounce the use of lethal biological agents and weapons, and all other methods of biological warfare. The United States will confine its biological research to defensive measures such as immunization and safety measures.

In his speech Nixon called his move "unprecedented"; and it was in fact the first review of the U.S. biological warfare program since 1954. Despite the lack of review, the biological warfare program had increased in cost and size since 1961; when Nixon ended the program the budget was $300 million annually. Nixon's statement confined all biological weapons research to defensive-only and ordered the destruction of the existing U.S. biological arsenal.

U.S. biological weapons stocks were destroyed over the next few years. A $12 million disposal plan was undertaken at Pine Bluff Arsenal, where all U.S. anti-personnel biological agents were stored. That plan was completed in May 1972 and included decontamination of facilities at Pine Bluff. Other agents, including anti-crop agents such as wheat stem rust, were stored at Beale Air Force Base and Rocky Mountain Arsenal. These anti-crop agents, along with agents at Fort Detrick used for research purposes were destroyed in March 1973.

Budget History

From the onset of the U.S. biological weapons program in 1943 through the end of World War II the United States spent $400 million on biological weapons, mostly on research and development. The budget for fiscal year 1966 was $38 million. When Nixon ended the U.S. bio-weapons program it represented the first review of the U.S. biological warfare program since 1954. Despite the lack of review, the biological warfare program had increased in cost and size since 1961; when Nixon ended the program the budget was $300 million annually.

Geneva Protocol and BWC

The 1925 Geneva Protocol, ratified by most major powers in the 1920s and 30s, had still not been ratified by the United States at the dawn of World War II. Among the Protocol's provisions, was a ban on bacteriological warfare. The Geneva Protocol had encountered opposition in the U.S. Senate, in part due to strong lobbying against it by the Chemical Warfare Service, and it was never brought to the floor for a vote when originally introduced. Regardless, on June 8, 1943 President Roosevelt affirmed a no-first-use policy for the United States concerning biological weapons. Even with Roosevelt's declaration opposition to the Protocol remained strong; in 1949 the Protocol was among several old treaties returned to President Harry S. Truman unratified.

When Nixon ended the U.S. bio-weapons program in 1969 he also announced that he would resubmit the Geneva Protocol to the U.S. Senate. This was a move Nixon was considering as early as July 1969. The announcement included language that indicated the Nixon administration was moving toward an international agreement on an outright ban on bio-weapons. Thus, the Nixon administration became the world's leading anti-biological weapons voice calling for an international treaty. The Eighteen Nation Disarmament Committee was discussing a British draft of a biological weapons treaty which the United Nations General Assembly approved in 1968 and that NATO supported. These arms control talks would eventually lead to the Biological Weapons Convention, the international treaty outlawing biological warfare. Prior, to the Nixon announcement only Canada supported the British draft. Beginning in 1972, the Soviet Union, United States and more than 100 other countries signed the BWC. The United States ratified the Geneva Protocol in 1975.

Post-1969 bio-defence Program

Both the U.S. bio-weapons ban and the Biological Weapons Convention restricted any work in the area of biological warfare to defensive in nature. In reality, this gives BWC member-states wide latitude to conduct biological weapons research because the BWC contains no provisions for monitoring of enforcement. The treaty, essentially, is a gentlemen's agreement amongst members backed by the long-prevailing thought that biological warfare should not be used in battle.

After Nixon declared an end to the U.S. bio-weapons program debate in the Army centered around whether or not toxin weapons were included in the president's declaration. Following Nixon's November 1969 order, scientists at Fort Detrick worked on one toxin, *Staphylococcus* enterotoxin type B (SEB), for several more months. Nixon ended the debate when he added toxins to the bio-weapons ban in February 1970. The U.S. also ran a series of experiments with anthrax, code named Project Bacchus, Project Clear Vision and Project Jefferson in the late 1990s and early 2000s.

In recent years certain critics have claimed the U.S. stance on biological warfare and the use of biological agents has differed from historical interpretations of the BWC. For example, it is said that the U.S. now maintains that the Article I of the BWC (which explicitly bans bio-weapons), does not apply to "non-lethal" biological agents. Previous interpretation was stated to be in line with a definition laid out in Public Law 101-298, the Biological Weapons Anti-Terrorism Act of 1989. That law defined a biological agent as:

any micro-organism, virus, infectious substance, or biological product that may be engineered as a result of biotechnology, or any naturally occurring or bioengineered component of any such microorganism, virus, infectious substance, or biological product, capable of causing death, disease, or other biological malfunction in a human, an animal, a plant, or another living organism; deterioration of food, water, equipment, supplies, or material of any kind...

According to the Federation of American Scientists, U.S. work on non-lethal agents exceeds limitations in the BWC.

Agents and Weapons

When the U.S. biological warfare program ended in 1969 it had developed seven mass-produced, battle-ready biological weapons in the form of agents that cause: anthrax, tularemia, brucellosis, Q-fever, VEE, and botulism. In addition Staphylococcal Enterotoxin B was produced as an incapacitating agent. In addition to the agents that were ready to be used the U.S. program conducted research into the weaponization of more than 20 other agents. They included: smallpox, EEE and WEE, AHF, Hantavirus, BHF, Lassa fever, glanders, melioidosis, plague, yellow fever, psittacosis, typhus, dengue fever, Rift Valley fever (RVF), CHIKV, late blight of potato, rinderpest, Newcastle disease, bird flu, and the toxin ricin.

Besides the numerous pathogens that afflict human beings, the U.S. had developed an arsenal of anti-agriculture biological agents. These included rye stem rust spores (stored at Edgewood Arsenal, 1951–1957), wheat stem rust spores (stored at the same facility 1962-1969), and the causative agent of rice blast.

A U.S. facility at Fort Terry primarily focused on anti-animal biological agents. The first agent that was a candidate for development was foot and mouth disease (FMD). Besides FMD, five other top secret biological weapons projects were commissioned on Plum Island. The other four programs researched included RVF, rinderpest, African swine fever, plus eleven miscellaneous exotic animal diseases. The eleven miscellaneous pathogens were: Blue tongue virus, bovine influenza, bovine virus diarrhea (BVD), fowl plague, goat pneumonitis, mycobacteria, "N" virus, Newcastle disease, sheep pox, Teschers disease, and vesicular stomatitis.

Work on delivery systems for the U.S. bio-weapons arsenal led to the first mass-produced biological weapon in 1952, the M33 cluster bomb. The M33's sub-munition, the pipe bomb like, cylindrical M114 bomb, was also completed and battle-ready by 1952. Other delivery systems researched and at least partially developed during the 1950s included the E77 balloon bomb and the E86 cluster bomb. The peak of U.S. biological weapons delivery system development came during the 1960s. Production of cluster bomb sub-muntions began to shift from the cylindrical bomblets to spherical bomblets, which had a larger coverage area. Development of the spherical E120 bomblet took place in the early 1960s as did development of the M143 bomblet, similar to the chemical M139 bomblet. The experimental Flettner rotor bomblet was also developed during this time period. The Flettner rotor was called, "probably one of the better devices for disseminating microorganisms", by William C. Patrick III.

Alleged Uses

Cuba

It has been rumored that the U.S. employed biological weapons against the Communist island nation of Cuba. Noam Chomsky claimed that evidence exists implicating the U.S. in biological warfare in

Cuba, but his evidence was disputed. Allegations in 1962 held that CIA operatives had contaminated a shipment of sugar while it was in storage in Cuba. Again, in 1962, a Canadian agricultural technician assisting the Cuban government claimed he was paid $5,000 to infect Cuban turkeys with the deadly Newcastle disease. Though the technician later claimed he had just pocketed the money, many Cubans and some Americans believed a clandestinely administered biological weapons agent was responsible for a subsequent outbreak of the disease in Cuban turkeys. In 1971 the first serious outbreak of swine flu in the Western Hemisphere occurred in Cuba, and Cubans alleged that U.S. covert biological warfare was responsible for this outbreak, which led to the preemptive slaughter of 500,000 pigs. Evidence linking these incidents to biological warfare has not been confirmed.

Accusations have continued to come out of Havana charging U.S. use of bio-weapons on the island. The Cuban government blamed the U.S. for a 1981 outbreak of dengue fever that sickened more than 300,000. Dengue, a vector-borne disease usually carried by mosquitoes, killed 158 people that year in Cuba, including 101 children under 15. Poor relations between Cuba and the United States coupled with confirmed U.S. research into entomological warfare during the 1950s made these charges seem not implausible. However, dengue fever occurs naturally in the region of the world where Cuba is located.

Korean War

North Korean and Chinese officials leveled accusations that during the Korean War the United States engaged in biological warfare in North Korea. The claim is dated to the period of the war, and has been thoroughly denied by the U.S. In 1998, Stephen Endicott and Edward Hagermann claimed that the accusations were true in their book, *The United States and Biological Warfare: Secrets from the Early Cold War and Korea* The book received mixed reviews, some called it "bad history" and "appalling", while other praised the case the authors made.

In 1952 the Chinese and North Koreans insinuated that mysterious outbreaks of disease in North Korea and China were due to U.S. biological attacks. Despite assertions that this did not occur from the International Red Cross and World Health Organization, whom the Chinese denounced as Western-biased, the Chinese government pursued an investigation by the World Peace Council. A committee led by Joseph Needham gathered evidence for a report that included eyewitness testimony, and testimony from doctors as well as four American Korean War prisoners who confirmed use of biological weapons by the U.S. The U.S. government denied the accusations and their denial was generally supported by top scientists in the West. In Eastern Europe, and China, North Korea it was widely believed that the accusations were true.

The same year Endicotts' book was published Kathryn Weathersby and Milton Leitenberg of the Cold War International History Project at the Woodrow Wilson Centre in Washington released a cache of Soviet and Chinese documents which revealed the North Korean claim was an elaborate disinformation campaign. In addition, a Japanese journalist claims to have seen similar evidence of a Soviet disinformation campaign and that the evidence supporting its occurrence was faked.

Others have revived these claims more recently. In March 2010, the allegations were investigated by the Al Jazeera English news programme People & Power. In this program, Professor Mori Masataka investigated historical artifacts in the form of bomb casings from US biological weapons, contemporary documentary evidence and eye witness testimonies. From the evidence he collected, Professor Mori concluded that the United States did in fact test biological weapons on North Korea during the Korean War.

Experimentation and Testing

Entomological Testing

The United States seriously researched the potential of entomological warfare (EW) during the Cold War. EW is a specific type of biological warfare which aims to use insects as weapon, either directly or through their potential to act as vectors. During the 1950s the United States conducted a series of field tests using entomological weapons. Operation Big Itch, in 1954, was designed to test munitions loaded with uninfected fleas (*Xenopsylla cheopis*). In May 1955 over 300,000 yellow fever mosquitoes (*Aedes aegypti*) were dropped over parts of the U.S. state of Georgia to determine if the air-dropped mosquitoes could survive to take meals from humans. The mosquito tests were known as Operation Big Buzz. The U.S. engaged in at least two other EW testing programs, Operation Drop Kick and Operation May Day. A 1981 Army report outlined these tests as well as multiple cost-associated issues that occurred with EW.

Experiments on Consenting Individuals

Operation Whitecoat involved the controlled testing of many serious agents on military personnel consented to experimentation, and understood the risks involved. No deaths are known to have resulted from this program.

Experiments on Non-consenting Individuals

Testing on Unwitting Military Personnel

In August 1949 a U.S. Army Special Operations Division, operating out of Fort Detrick in Maryland, set up its first test at The Pentagon in Washington, D.C. Operatives sprayed harmless bacteria into the building's air conditioning system and observed as the microbes spread throughout the Pentagon.

The US military acknowledges that it tested several chemical and biological weapons on US military personnel in the desert facility, including the East Demilitarization Area near Deseret Chemical Depot/ Deseret Chemical Test Centre at Fort Douglas, Utah, but takes the position that the tests have contributed to long-term illnesses in only a handful of exposed personnel. Veterans who took part believe they were also exposed to Agent Orange. The Department of Veterans Affairs denies almost all claims for care and compensation made by veterans who believe they got sick as a result of the tests. The US military for decades remained silent about "Project 112" and its victims, a slew of tests overseen by the Army's Deseret Test Centre in Salt Lake City. Project 112 starting in the 1960s tested chemical and biological agents, including VX, sarin and e. Coli, on military personnel who did not know they were being tested. After the Defence Department finally acknowledged conducting the tests on unwitting human subjects, it agreed to help the Veterans' Affairs Department track down those who were exposed, but a General Accountability Office report in 2008 scolded the military for ceasing the effort.

Testing on Unwitting Civilians

Medical experiments were conducted on a large scale on civilians who had not consented to participate. Often, these experiments took place in urban areas in order to test dispersion methods. Questions were raised about detrimental health effects after experiments in San Francisco, California, were followed by a spike in hospital visits; however, in 1977 the Centres for Disease Control and Prevention determined that there was no association between the testing and the occurrence of pneumonia or influenza. The San Francisco test involved a U.S. Navy ship that sprayed Serratia marcescens from the bay; it traveled more than 30 miles. One dispersion test involved laboratory personnel disguised as passengers spraying harmless bacteria in Ronald Reagan Washington National Airport.

Scientists tested biological pathogens, including *Bacillus globigii*, which were thought to be harmless, at public places such as subways. A light bulb containing *Bacillus globigii* was dropped on New York City's subway system; the result was strong enough to affect people prone to illness (also known as Subway Experiment). Based on the circulation measurements, thousands of people would have been killed if a dangerous microbe was released in the same manner. A jet aircraft released material over Victoria, Texas, that was monitored in the Florida Keys.

GAO Report

In February, 2008, the Government Accountability Office (GAO) released report GAO-08-366 titled, "Chemical and Biological Defence, DOD and VA Need to Improve Efforts to Identify and Notify Individuals Potentially Exposed during Chemical and Biological Tests." The report stated that tens of thousands of military personnel and civilians may have been exposed to biological and chemical substances through DOD tests. In 2003, the DOD reported it had identified 5,842 military personnel and estimated 350 civilians as being potentially exposed during the testing, known as Project 112.

The GAO scolded the U.S. Department of Defence's (DOD) 2003 decision to stop searching for people affected by the tests was premature. The GAO report also found that the DoD made no effort to inform civilians of exposure, and that the United States Department of Veterans Affairs (VA) is failing to use available resources to inform veterans of possible exposure or to determine if they were deceased. After the DoD halted efforts to find those who may have been affected by the tests, veteran health activisits and others identified approximately 600 additional individuals who were potentially exposed during Project 112.

Some of the individuals were identified after the GAO reviewed records stored at the Dugway Proving Ground, others were identified by the Institute of Medicine. Many of the newly identified suffer from long term illnesses that may have been caused by the biological or chemical testing.

- Fort Detrick, Maryland :
  - o U.S. Army Biological Warfare Laboratories (1943–69)
    - * Building 470
    - * One-Million-Liter Test Sphere
    - * Operation Whitecoat
  - o United States Army Medical Unit (1954–69)
  - o U.S. Army Medical Research Institute of Infectious Diseases (USAMRIID; 1969–present)
  - o National Biodefence Analysis and Countermeasures Centre (NBACC; Projected: 2008)
- Project Bacchus
- Project Clear Vision
- Project SHAD.

Project SHAD stands for Project Shipboard Hazard and Defence, a series of Cold War-era tests by the United States Department of Defence of biological weapons and chemical weapons. Exposures of uninformed and unwilling humans during the testing to the test substances, particularly the exposure to United States military personnel then in service, has added controversy to recent revelations of the project.

History

Project SHAD was part of a larger effort by the Department of Defence called Project 112. The Project began in 1962 during John F. Kennedy's administration, and it is largely believed that neither Kennedy nor subsequent Presidents knew of Project 112 or SHAD. However, Robert McNamara, Kennedy's Secretary of Defence, did know of and approved these tests.

There is also some evidence that demonstrates local governments were involved with these tests, though it is unclear how exactly they aided with Project SHAD. The official statement on Project SHAD's purpose was "...to identify U.S. war ships vulnerabilities to attacks with biological or chemical warfare agents and to develop procedures to respond to such attacks while maintaining a warfighting capability." 134 tests were planned initially, but only 46 tests were actually completed. In these tests, chemical and biological agents were introduced to military personnel, who were at the time ignorant that they were involved in such an experiment. Nerve agents and chemicals include, but are not limited to, VX nerve gas, Tabun gas, Sarin, Soman, and the marker chemicals zinc sulfide, cadmium sulfide, and QNB. Biologics include *Bacillus globigii*, *Coxiella burnetti* (which causes Q fever), and Francisella tularensis (which causes tularemia or 'rabbit fever').

Declassification

Revelations concerning Project SHAD were first exposed by Independent Producer and Investigative Journalist Eric Longabardi of TeleMedia News Productions, now based in Los Angeles, CA. Longabardi's 6-year investigation into the still secret program began in early 1994. It ultimately resulted in a series of investigative reports produced by him, which were broadcast on the CBS Evening News in May 2000. After the broadcast of these exclusive reports, the Pentagon and Veteran's Administration opened their own on going investigations into the long classified program. In 2002, Congressional hearings on Project SHAD, in both the Senate and House, further shed media attention on the still classified program. In 2002, a class action federal lawsuit was filed on behalf of the US Navy sailors exposed in the testing. Additional actions, including a multi-year medical study was conducted by National Academy of Sciences/Institute of Medicine to assess the potential medical harm caused to the thousands of unwitting US Navy sailors, civilians, and others who were exposed in the secret testing. The results of that study were finally released in May 2007.

28 fact sheets have been released, focusing on the Deseret Test Centre in Dugway, Utah, which was built entirely for Project SHAD and was closed after the Project was finished in 1973.

The US Department of Defence (DoD) has come under great scrutiny because those that were involved with Project 112 and SHAD were unaware of any tests being done. No effort was made to ensure the informed consent of the military personnel. Until 1998, the Department of Defence stated officially that Project SHAD did not exist. Because the DoD refused to acknowledge the program, surviving test subjects have been unable to obtain disability payments for health issues related to the project. US Representative Mike Thompson said of the program and the DoD's effort to conceal it, "They told me – they said, but don't worry about it, we only used simulants. And my first thought was, well, you've lied to these guys for 40 years, you've lied to me for a couple of years. It would be a real leap of faith for me to believe that now you're telling me the truth." The Department of Veterans Affairs has commenced a three-year study comparing known SHAD-affected veterans to veterans of similar ages who were not involved in any way with SHAD or Project 112. The study cost approximately US$3 million, and results are being compiled for future release.

United Kingdom

The United Kingdom has possessed weapons of mass destruction, including nuclear, biological, and chemical weapons. The United Kingdom is one of the five official nuclear weapon states under the Nuclear Non-Proliferation Treaty and has an independent nuclear deterrent. The U.K. has been estimated to have a stockpile of approximately 160 active nuclear warheads and 225 nuclear warheads in total. The United Kingdom renounced the use of chemical and biological weapons in 1956 and subsequently destroyed its general stocks.

Biological Weapons

During the Second World War, British scientists studied the use of biological weapons, including a test using anthrax on the Scottish island of Gruinard which left it contaminated and fenced off for nearly fifty years, until an intensive four-year programme to eradicate the spores was completed in 1990. They also manufactured five million linseed-oil cattle cakes with a hole bored into them for addition of anthrax spores between 1942 and mid-1943.

These were to be dropped on Germany using specially designed containers each holding 400 cakes, in a project known as Operation Vegetarian. It was intended that the disease would destroy the German beef and dairy herds and possibly spread to the human population. Preparations were not complete until early 1944. Operation Vegetarian was only to be used in the event of a German anthrax attack on Britain.

Offensive weapons development continued after the war into the 1950s with tests of plague, brucellosis, tularemia and later equine encephalomyelitis and vaccinia viruses (the latter as a relatively safe simulant for smallpox).

In particular five sets of trials took place at sea using aerosol clouds and animals.

- Operation Harness off Antigua in 1948-1949.

- Operation Cauldron off Stornoway in 1952. The trawler *Carella* accidentally sailed through a cloud of pneumonic plague bacilli (*yersinia pestis*) during this trial. It was kept under covert observation until the incubation period had elapsed but none of the crew fell ill.
- Operation Hesperus off Stornoway in 1953.
- Operation Ozone off Nassau in 1954.
- Operation Negation off Nassau in 1954-5.

The programme was cancelled in 1956 when the British government renounced the use of biological and chemical weapons. It ratified the Biological and Toxin Weapons Convention in March 1975.

- Porton Down
- Gruinard Island
- Nancekuke.

Former Soviet Union and Russia

The Soviet Union began a biological weapons program in the 1920s at the Leningrad Military Academy in Moscow under the control of the state security apparatus, known as the GPU. This occurred despite the fact that the USSR was a signatory to the 1925 Geneva Convention, which banned both chemical and biological weapons.

During World War II Stalin was forced to move his biological warfare operations out of the path of advancing German forces At the conclusion of the war, Soviet troops invading Manchuria captured many Unit 731 Japanese scientists and learned of their extensive human experimentation through captured documents and prisoner interrogations. Emboldened by these discoveries, Stalin put KGB chief Lavrenty Beria in charge of a new biowarfare program.

By 1960, numerous biological warfare research facilities existed throughout the Soviet Union. Although the USSR also signed the 1972 Biological Weapons Convention (BWC), the Soviets subsequently augmented their biowarfare programs. They doubted the United States' claimed compliance with the BWC, which further motivated their program. The Soviet biological weapons effort became a huge program, comprising various institutions under different ministries along with commercial facilities and collectively known as Biopreparat after 1973. Biopreparat pursued offensive research, development, and production of biological agents under cover of legitimate civil biotechnology research. It conducted its clandestine activities at 52 sites and employed over 50,000 people. Annualized production capacity for weaponized smallpox, for example, was 90 to 100 tons.

In the 1990s, President of the Russian Federation Boris Yeltsin admitted to an offensive bio-weapons program as well as to the true nature of the Sverdlovsk biological weapons accident of 1979, which had resulted in the deaths of at least 64 people Soviet defectors, including Colonel Kanatjan Alibekov, first deputy chief of Biopreparat from 1988 to 1992, confirmed that the program had been massive and still existed. In September 1992, Russia signed an agreement with the United States and Great Britain promising to end its bio-weapons program and to convert its facilities for benevolent scientific and medical purposes. Compliance with the agreement as well as the fate of the former Soviet bio-agents and facilities, is still mostly undocumented.

History

Timeline

1928 : Revolutionary Military Council signed a decree about weaponization of typhus. Leningrad Military academy began cultivation of typhus in chicken embryos. Human experimentation with typhus, glanders and melioidosis in Solovetsky camp. A laboratory on vaccine and serum research was also established near Moscow in 1928, within Military Chemical Agency. This laboratory was transformed to Red Army's Scientific Research Institute of Microbiology in 1933.

1941: Soviet bioweapons facilities are evacuated to the city of Kirov.

1942: Alleged use of tularemia against German troops.

1945: Japanese documentation from Unit 731 was captured.

1946: A biological weapons facility was established in Sverdlovsk.

1953: Fifteenth directorate of Red Army takes responsibility for the program.

1973: A "civilian" main directorate Biopreparat was founded. Other organizations involved in design and production of biological weapons were Soviet Ministry of Defence, Ministry of Agriculture., Ministry of Health., USSR Academy of Sciences, and KGB.

1990s: specimens of deadly bacteria and viruses were stolen from Western laboratories and delivered by Aeroflot planes to support

Russian program of biological weapons. At least one of the pilots was a Russian Foreign Intelligence Service officer". At least two agents died, presumably from the transported pathogens. Beginning of 2000s: Academician "A.S." proposed new biological warfare program "Biological Shield of Russia" to president Vladimir Putin. The program reportedly includes institutes of Russian Academy of Sciences from Pushchino.

Military use During World War II

Tularemia was allegedly used against German troops in 1942 near Stalingrad. Around 10,000 cases of tularemia had been reported in the Soviet Union during the years of 1941 and 1943. However, the number of cases jumped to more than 100,000 in the year of Stalingrad outbreak. German panzer troops fell ill in such significant numbers during the late summer of 1942 that German military campaign came to a temporary halt. German soldiers became ill with the rare pulmonary form of tularemia, which indicate the use of an aerosol biological weapon (the ordinary transmission pathway is through ticks and rodents). According to Kenneth Alibek the used tularemia weapon had been developed in the Kirov military facility. It was suggested by some, however, that the outbreak might be of natural origin, since a pulmonary form of tularemia has also been noted in natural outbreaks in Martha's Vineyard in 2000.

In the Soviet Union the outbreak at Stalingrad was described as a natural outbreak. Crops were left in the field during the German offensive and the rodent population swelled putting many inhabitants into contact with infected rodents. In some parts of the Stalingrad Oblast as many as 75% of the inhabitants became infected. It was noted that before the war there was a so-called threshing tularemia caused by people inhaling infected dusts soiled by rodents while threshing grain.

Post-BWC Developments

Soviet Union continued development and mass production of offensive biological weapons, despite having signed the 1972 Biological Weapons Convention (BWC). The development and production was conducted by main directorate "Biopreparat", Soviet Ministry of Defence, Ministry of Agriculture, Ministry of Health, USSR Academy of Sciences, the KGB, and other state organizations.

In 1980s Soviet Ministry of Agriculture had successfully developed variants of foot-and-mouth disease and rinderpest against cows, African swine fever for pigs, and psittacosis to kill chicken. These agents were prepared to spray them down from tanks attached to airplanes over hundreds of miles. The secret program was code-named "Ecology".

Notable Outbreaks and Accidents

Marburg Virus

The Soviet Union reportedly had a large biological weapons program enhancing the usefulness of the Marburg virus. The development was conducted in Vector Institute under leadership of Dr. Ustinov who accidentally died from the virus. The samples of Marburg taken from Ustinov's organs were more powerful than the original strain. New strain called "Variant U" had been successfully weaponized and approved by Soviet Ministry of Defence in 1990.

Smallpox

The first smallpox weapons factory in the Soviet Union was established in 1947 in the city of Zagorsk, close to Moscow. It was produced by injecting small amounts of the virus into chicken eggs. An especially virulent strain was brought from India in 1967 by a special Soviet medical team that was sent to India to help to eradicate the virus. The pathogen was manufactured and stockpiled in large quantities throughout the 1970s and 1980s.

An outbreak of weaponized smallpox occurred during its testing in the 1970s. General Prof. Peter Burgasov, former Chief Sanitary Physician of the Soviet Army, and a senior researcher within the program of biological weapons described this incident:

> *"On Vozrozhdeniya Island in the Aral Sea, the strongest recipes of smallpox were tested. Suddenly I was informed that there were mysterious cases of mortalities in Aralsk. A research ship of the Aral fleet came 15 km away from the island (it was forbidden to come any closer than 40 km). The lab technician of this ship took samples of plankton twice a day from the top deck. The smallpox formulation— 400 gr. of which was exploded on the island—"got her" and she became infected. After returning home to Aralsk, she infected several people including children. All of them died. I suspected the reason for this and called the Chief of General Staff of Ministry of Defence and requested to forbid the stop of the Alma-Ata to Moscow train in Aralsk. As a result, the epidemic around the country was prevented. I called*

Andropov, who at that time was Chief of KGB, and
informed him of the exclusive recipe of smallpox obtained
on Vozrozhdeniya Island."

A production line to manufacture smallpox on an industrial scale was launched in the Vector Institute in 1990. The development of genetically altered strains of smallpox was presumably conducted in the Institute under leadership of Dr. Sergei Netyosov in the middle of the 1990s, according to Kenneth Alibek.

It was reported that Russia made smallpox available to Iraq in the beginning of 1990s.

Anthrax

Spores of weaponized anthrax were accidentally released from a military facility near the city of Sverdlovsk in 1979. The death toll was at least 105, but no one knows the exact number, because all hospital records and other evidence were destroyed by the KGB, according to former Biopreparat deputy director Kenneth Alibek.

Sverdlovsk Anthrax Leak

The Sverdlovsk anthrax leak is an incident when spores of anthrax were accidentally released from a military facility in the city of Sverdlovsk (formerly, and now again, Yekaterinburg) 1450 km east of Moscow on April 2, 1979. This accident is sometimes called "biological Chernobyl". The ensuing outbreak of the disease resulted in approximately 100 deaths, although the exact number of victims remains unknown. The cause of the outbreak had for years been denied by the Soviet Union, which blamed the deaths on intestinal exposure due to the consumption of tainted meat from the area, and subcutaneous exposure due to butchers handling the tainted meat. All medical records of the victims had been removed in order to avoid revelations of serious violations of the Biological Weapons Convention.

Background

The closed city of Sverdlovsk had been a major production centre of the Soviet military-industrial complex since World War II. It produced tanks, nuclear rockets and other armaments. A major nuclear accident happened in this region in 1958, when a military reactor was damaged, resulting in the spread of radioactive dust over a thousand square kilometres. The biological weapons facility in Sverdlovsk was built after World War II, using documentation captured in Manchuria from the Japanese germ warfare program.

The strain of anthrax produced in *Military Compound 19* near Sverdlovsk was the most powerful in the Soviet arsenal ("Anthrax 836"). It had been isolated as a result of another anthrax leak accident that happened in 1953 in the city of Kirov. A leak from a bacteriological facility contaminated the city sewer system. In 1956, biologist Vladimir Sizov found a more virulent strain in rodents captured in this area. This strain was planned to be used to arm warheads for the SS-18 ICBM, which would target American cities, among other targets.

The Accident

The produced anthrax culture had to be dried to produce a fine powder for use as an aerosol. Large filters over the exhaust pipes were the only barriers between the anthrax dust and the outside environment. On the last Friday of March 1979, a technician removed a clogged filter while drying machines were temporarily turned off. He left a written notice, but did not write this down in the logbook as he was supposed to do. The supervisor of the next shift did not find anything unusual in the logbook, and turned the machines on. In a few hours, someone found that the filter was missing and reinstalled it. The incident was reported to military command, but local and city officials were not immediately informed. Boris Yeltsin, a local Communist Party boss at this time, was helping to cover up the accident. All workers of a ceramic plant across the street fell ill during the next few days. Almost all of them died in a week. The death toll was at least 105, but the exact number is unknown as all hospital records and other evidence were destroyed by the KGB, according to former Biopreparat deputy director Ken Alibek.

The Investigation

In the 1980s, there was vigorous international debate and speculation as to whether the outbreak was natural or an accidental exposure. If accidental, there was discussion of whether it represented violation of the 1972 Biological Weapons Convention. A number of small investigations launched by Russian scientists in the years immediately following the dissolution of the Soviet Union re-opened the case in a number of newspaper articles. A team of Western inspectors lead by Professor Matthew Meselson of Harvard finally gained access to the region in 1992, and determined that all of the victims had been living directly downwind at the time of the release of the spores via aerosol. Livestock in the area were also affected. It was revealed around this time that the accident was caused by the non-replacement of a filter on an exhaust at the facility, and though the problem was quickly rectified it was too late to prevent a release. Had the winds been blowing in the direction of the city at that time, it could have resulted in the pathogen being spread to hundreds of thousands of people. The military facility remains closed to inspection. Professor Meselson's original contention for many years had been that the outbreak was a natural one and that the Soviet authorities were not lying when they disclaimed having an active offensive bio-warfare program, but the information uncovered in the investigation left no room for doubt. Meselson's wife, Jeanne Guillemin (who had participated in the investigation), detailed the events in a 1999 book.

Aftermath

Russian Prime Minister Egor Gaidar issued a decree to begin demilitarization of Compound 19 in 1992. However, the facility continued its work. Not a single journalist has been allowed onto the premises since 1992. About 200 soldiers with Rottweiler dogs still patrol the complex. Classified activities were moved underground, and several new laboratories have been constructed and equipped to work with highly dangerous pathogens. One of their current subjects is reportedly *Bacillus anthracis* strain H-4. Its virulence and antibiotic resistance have been dramatically increased using genetic engineering.

- Biopreparat (18 labs and production centres)
  - Stepnagorsk Scientific and Technical Institute for Microbiology, Stepnogorsk, northern Kazakhstan
  - Institute of Ultra Pure Biochemical Preparations, Leningrad, a weaponized plague centre
  - Vector State Research Centre of Virology and Biotechnology (VECTOR), a weaponized smallpox centre
  - Institute of Applied Biochemistry, Omutninsk
  - Kirov bioweapons production facility, Kirov, Kirov Oblast
  - Zagorsk smallpox production facility, Zagorsk
  - Berdsk bioweapons production facility, Berdsk
  - Sverdlovsk bioweapons production facility (Military Compound 19), Sverdlovsk, a weaponized anthrax centre.

Poison Laboratory of the Soviet Secret Services

Poison laboratory of the Soviet secret services, alternatively known as Laboratory 1, Laboratory 12, and *Kamera* which means "The

Chamber" in Russian, was a covert poison research and development facility of the Soviet secret police agencies.

Chronology

- 1921: First poison laboratory within the Soviet secret services was established under the name "Special Office". It was headed by professor of medicine Ignatii Kazakov, according to Pavel Sudoplatov.

- 1926: The laboratory was under the supervision of Genrikh Yagoda, a deputy of OGPU chairman Vyacheslav Menzhinsky, who became NKVD chief in 1934 after Menzhinsky's death.

- February 20, 1939: It becomes *Laboratory 1* headed by Grigory Mairanovsky. The laboratory was under the direct supervision of NKVD director Lavrenty Beria and his deputy Vsevolod Merkulov from 1939 to March 1953.

- December 21, 1951: Grigory Mairanovsky arrested in connection with Viktor Abakumov's arrest, which was presumably a part of Joseph Stalin's campaign to remove NKVD chief, Lavrenty Beria.

- March 14, 1953: It was renamed to *Laboratory 12*. V. Naumov is the newly appointed head. Lavrenty Beria and Vsevolod Merkulov were executed after Stalin's death. Immediate NKVD supervisor of the laboratory, Pavel Sudoplatov, received long prison sentences.

- 1978: Expanded into the *Central Investigation Institute for Special Technology* within the First Chief Directorate of the KGB

- Currently: Several laboratories of the SVR, (headquartered in Yasenevo near Moscow), are responsible for the "creation of biological and toxin weapons for clandestine operations in the West".

Human Experimentation

Mairanovsky and his colleagues tested a number of deadly poisons on prisoners from the Gulag ("enemies of the people"), including mustard gas, ricin, digitoxin and many others. The goal of the experiments was to find a tasteless, odourless chemical that could not be detected *post mortem*. Candidate poisons were given to the victims, with a meal or drink, as "medication". Finally, a preparation with the desired properties called C-2 was developed According to witness testimonies, the victim changed physically, became shorter, weakened quickly, became calm and silent and died within fifteen minutes. Mairanovsky brought to the laboratory people of varied physical condition and ages in order to have a more complete picture about the action of each poison.

Pavel Sudoplatov and Nahum Eitingon approved special equipment [poisons] only if it had been tested on humans", according to testimony of Mikhail Filimonov. Vsevolod Merkulov said that these experiments were approved by NKVD chief Lavrenty Beria.. Beria himself testified on August 28, 1953, after his arrest that "I gave orders to Mairanovsky to conduct experiments on people sentenced to the highest measure of punishment, but it was not my idea".

In addition to human experimentation, Mairanovsky personally executed people with poisons, under the supervision of Sudoplatov

Prominent Victims

- The leader of the Russian All-Military Union general Alexander Kutepov was drugged and kidnapped in Paris in 1930. He died from a heart attack due to an overdose of the administered drug.

- One of leaders of the White movement, Russian general Evgenii Miller, was drugged and kidnapped in Paris in 1937. He was executed later in Russia.

- Abram Slutsky, head the Soviet foreign intelligence service (GUGB) was poisoned with hydrocyanic acid added in tea in 1938

- Archbishop Theodore Romzha of Ukrainian Catholic Church was killed in 1947 by injection of curare provided by Mairanovsky and administered by a medical nurse who was an Ministry for State Security (USSR) agent.

- In 1978, dissident Bulgarian writer Georgi Markov was assassinated in London using a tiny pellet poisoned with ricin; the necessary equipment was prepared in this laboratory. In a Discovery Channel television program about his illustrated book of espionage equipment called "The Ultimate Spy," espionage historian H. Keith Melton indicated that once the Bulgarian secret police had decided to kill Markhov, KGB specialists from the Laboratory gave the Bulgarians a choice between two KGB tools that could be provided for the task — either a poisonous topical gelatin to be smeared on Markhov, or an instrument to administer a poison pellet, as was eventually done.

• Attempted poisoning of the second President of Afghanistan Hafizullah Amin on December 13, 1979. Department 8 of KGB succeeded in infiltrating the illegal agent Mitalin Talybov (codenamed SABIR) as a chef of Amin's presidential palace. However, Amin switched his food and drink as if he expected to be poisoned, so his son-in-law became seriously ill, and ironically, was flown to a hospital in Moscow.

Alleged Victims

• Russian writer Maksim Gorky and his son. During the Trial of the Twenty One in 1938, NKVD chief Genrikh Yagoda admitted that he poisoned to death Maksim Gorky and his son and unsuccessfully tried to poison future NKVD boss Nikolay Yezhov. The attempted poisoning of Yezhov was later officially dismissed as falsification, but Vyacheslav Molotov believed that the poisoning accusations were true. Yagoda was never officially rehabilitated (recognized as an innocent victim of political repressions) by Soviet authorities.

• Soviet leader Joseph Stalin. Russian historians Anton Antonov-Ovseenko and Edvard Radzinsky believe that Stalin was poisoned by associates of NKVD chief Lavrentiy Beria, based on the interviews of a former Stalin body guard and numerous circumstantial evidence. Stalin planned to dismiss and execute Beria and other senior members of the Soviet government in 1953. According to Radzinsky, Stalin was poisoned by Khrustalev, a senior bodyguard briefly mentioned in memoirs of Svetlana Alliluyeva, Stalin's daughter.

• Journalist Anna Politkovskaya. During the Beslan school hostage crisis in September 2004 and while on her way to Beslan to help in negotiations with the hostage-takers, Politkovskaya fell violently ill and lost consciousness after drinking tea. She survived. According to a report by The Sunday Times, the drug was prepared in the FSB poison facility.

Planned Victims

• President of Socialist Federal Republic of Yugoslavia Josip Broz Tito. In the late 1940s, the laboratory manufactured a powdered plague for use in a small container and where the assassin was vaccinated against plague. The device was to be used against Tito, but MGB agent Iosif Grigulevich, who previously organized the assault on villa of Leon Trotsky and now have received the assignment to kill Tito, was recalled after the death of Stalin.

· The first democratically elected President of the Republic of Georgia, Zviad Gamsakhurdia. According to former Deputy Director of Biopreparat Ken Alibek, this laboratory was possibly involved in design of undetectable chemical or biological agent to assassinate Gamsakhurdia. BBC News reported that some Gamsakhurdia friends believed he committed suicide, "although his widow insists that he was murdered."

Vozrozhdeniya

Vozrozhdeniya, also known as Rebirth Island, was a former island of the Aral Sea or South Aral Sea. Due to the ongoing shrinkage of the Aral, it became first a peninsula in Mid 2001 and finally part of the mainland. Since the disappearance of the Southeast Aral in 2008, Vozrozhdeniya effectively no longer exists as a distinct geographical feature. The area is now shared by Kazakhstan and Uzbekistan.

Located in the central Aral Sea, Vozrozhdeniya Island was one of the main laboratories and testing sites for the Soviet Union's Microbiological Warfare Group. In 1948, a top-secret Soviet bioweapons laboratory was established here, which tested a variety of agents, including anthrax, smallpox, plague, brucellosis, and tularemia.

In the 1990s, word of the island's danger was spread by Soviet defectors, including Ken Alibek, the former head of the Soviet Union's bioweapons program. It was here, according to just released documents, that anthrax spores and bubonic plague bacilli were made into weapons and stored. The main town on the island was Kantubek, which lies in ruins today, but once had approximately 1,500 inhabitants.

The laboratory staff members abandoned the small island in 1992. Many of the containers holding the spores were not properly stored or destroyed, and over the last decade many of these containers have developed leaks.

In 2002, through a project organized by the United States and with Uzbekistan assistance, 10 anthrax burial sites were decontaminated. According to the Kazakh Scientific Centre for Quarantine and Zoonotic Infections, all burial sites of anthrax were decontaminated.

Part of the campaign in the 2010 video game Call of Duty: Black Ops takes place on Vozrozhdeniya Island.

Japan

Special Japanese military units conducted experiments on civilians and POWs in China. One of the most infamous was Unit 731 under Shirô Ishii. Victims were subjected to vivisection without anesthesia, amputations, and were used to test biological weapons, among other experiments. Anesthesia was not used because it was believed to affect results.

To determine the treatment of frostbite, prisoners were taken outside in freezing weather and left with exposed arms, periodically drenched with water until frozen solid. The arm was later amputated; the doctor would repeat the process on the victim's upper arm to the shoulder. After both arms were gone, the doctors moved on to the legs until only a head and torso remained. The victim was then used for plague and pathogens experiments.

According to GlobalSecurity.org, the experiments carried out by Unit 731 alone caused 3,000 deaths. Furthermore, according to the 2002 *International Symposium on the Crimes of Bacteriological Warfare*, the number of people killed by the Imperial Japanese Army germ warfare and human experiments is around 580,000. According to other sources, "tens of thousands, and perhaps as many as 400,000, Chinese died of bubonic plague, cholera, anthrax and other diseases...", resulting from the use of biological warfare.

One case of human experimentation occurred in Japan itself. At least nine out of 12 crew members survived the crash of a U.S. Army Air Forces B-29 bomber on Kyushu, on May 5, 1945. (This plane was Lt. Marvin Watkins' crew of the 29th Bomb Group of the 6th Bomb Squadron.) The bomber's commander was sent to Tokyo for interrogation, while the other survivors were taken to the anatomy department of Kyushu University, at Fukuoka, where they were subjected to vivisection or killed. On March 11, 1948, 30 people including several doctors were brought to trial by the Allied war crimes tribunal. Charges of cannibalism were dropped, but 23 people were found guilty of vivisection or wrongful removal of body parts. Five were sentenced to death, four to life imprisonment, and the rest to shorter terms. In 1950, the military governor of Japan, General Douglas MacArthur, commuted all of the death sentences and significantly reduced most of the prison terms. All of those convicted in relation to the university vivisection were free by 1958. In addition, many participants who were responsible for these vivisections were never charged by the Americans or their allies in exchange for the information on the experiments.

In 2006, former IJN medical officer Akira Makino stated that he was ordered—as part of his training—to carry out vivisection on about 30 civilian prisoners in the Philippines between December 1944 and February 1945. The surgery included amputations. Ken Yuasa, a former military doctor in China, has also admitted to similar incidents in which he was compelled to participate.

- Unit 731
- Zhongma Fortress
- Unit 100
- Unit 2646
- Unit 8604
- Unit Ei 1644.

Iraq

Saddam Hussein initiated an extensive biological weapons (BW) program in Iraq in the early 1980s, in violation of the Biological Weapons Convention (BWC) of 1972. Details of the BW program — along with a chemical weapons program — surfaced only in the wake of the Gulf War (1990–91) following investigations conducted by the United Nations Special Commission (UNSCOM) which had been charged with the post-war disarmament of Saddam's Iraq.

Because of this UN disarmament program, more is known today about the once-secret bioweapons program in Iraq than that of any other nation.

The Program

Startup and Foreign Suppliers

In the early 1980s, five German firms supplied equipment to manufacture botulin toxin and mycotoxin to Iraq. Iraq's State Establishment for Pesticide Production (SEPP) also ordered culture media and incubators from Germany's Water Engineering Trading.

Strains of dual-use biological material from France also helped advance Iraq's biological warfare program. From the United States, the non-profit American Type Culture Collection and the U.S. Centres for Disease Control sold or sent biological samples to Iraq up until 1989, which Iraq claimed to need for medical research. These materials included anthrax, West Nile virus and botulism, as well as *Brucella melitensis*, and *Clostridium perfringens*.

Some of these materials were used for Iraq's biological weapons research program, while others were used for vaccine development. In delievering these materials "The CDC was abiding by World Health Organization guidelines that encouraged the free exchange of biological samples among medical researchers..." according to Thomas Monath, CDC lab director. It was a request "which we were obligated to fulfill," as described in WHO and UN treaties.

Facilities, Agents and Production

Iraq's BW facilities included its main biowarfare research centre at Salman Pak (just south of Baghdad), the main bioweapons production facility at Al Hakum (the "Single-Cell Protein Production Plant") and the viral biowarfare research site at Al Manal (the "Foot and Mouth Disease Centre").

The Al Hakum facility began mass production of weapons-grade anthrax in 1989, eventually producing 8,000 litres or more (the 8,000 liter figure is based on declared amounts). Iraq officially acknowledged that it had worked with several species of bacterial pathogen, including *Bacillus anthracis*, *Clostridium botulinum* and *Clostridium perfringens* (gas gangrene) and several viruses (including enterovirus 17 (human conjunctivitis), rotavirus and camelpox).

The program also purified biological toxins, such as botulinum toxin, ricin and aflatoxin. After 1995, it was learned that, in all, Iraq had produced 19,000 litres of concentrated botulinum toxin (nearly 10,000 litres filled into munitions), 8,500 litres of concentrated anthrax (6,500 litres filled into munitions) and 2,200 litres of aflatoxin (1,580 litres filled into munitions). In total, the program grew a half million litres of biological agents.

- Al Hakum
- Salman Pak facility
- Al Manal facility.

Treaties Banning or Restricting BW

- Geneva Protocol
- Biological Weapons Convention.

List of People Associated with Biological Weapons

Bioweaponeers:

- Anton Dilger
- Ira Baldwin
- Paul Fildes
- Rihab Rashid Taha
- William C. Patrick III
- Kanatjan Alibekov, known as Ken Alibek
- Vladimir Pasechnik
- Kurt Blome
- Eugen von Haagen
- Kurt Gutzeit
- Erich Traub
- Shiro Ishii.

AIDS Origins Conspiracy Theories

AIDS conspiracy theories are claims or hypotheses about the origins and/or nature of human immunodeficiency virus (HIV) and AIDS that differ radically from the scientific consensus.

These alternative ideas range from suggestions that AIDS was the inadvertent result of experiments in the development of vaccines, to claims that HIV was developed by scientists working for the U.S. government.

While a few reputable mainstream scientists once investigated some of these theories as reasonable hypotheses, this is no longer the case, as continuing research has invalidated the alternative ideas. Recent evidence indicates that AIDS originated in Africa in the mid 1930s from the closely related Simian immunodeficiency virus, today found in monkeys and chimpanzees.

Man-made or Iatrogenic Origins of AIDS

- In an interview by Time magazine with Nobel Peace Prize laureate and environmental activist Wangari Maathai, it was alleged that Maathai had said that "AIDS is a biological weapon manufactured by the developed world to wipe out the black race". Maathai subsequently rejected that in a written statement issued in December 2004: "I neither say nor believe that the virus was developed by white people or white powers in order to destroy the African people. Such views are wicked and destructive."

- Jakob Segal, a former biology professor at Humboldt University in then-East Germany, proposed that HIV was engineered at a U.S. military laboratory at Fort Detrick, by splicing together two other viruses, Visna and HTLV-1. According to his theory, the new virus, created between 1977 and 1978, was tested on prison inmates who had volunteered for the experiment in exchange for early release. He further suggested that it was through these prisoners that the virus was spread to the population at large. He has been accused, however, by KGB defector Vasili Mitrokhin as having been disseminating disinformation on behalf of the Soviet Union, and the disease is known today to have existed in humans since at least 1959.

- Alan Cantwell, in self-published books entitled *AIDS and the Doctors of Death: An Inquiry into the Origin of the AIDS Epidemic* and *Queer Blood: The Secret AIDS Genocide Plot*, says that HIV is a genetically modified organism developed by U.S. Government scientists and that it was introduced into the population through Hepatitis B experiments performed on gay and bisexual men between 1978–1981 in major U.S. cities. Cantwell claims that these experiments were directed by Wolf Szmuness, and that there was an ongoing government cover-up of the origins of the AIDS epidemic. Similar theories have been advanced by Robert B. Strecker, Matilde Krim and Milton William Cooper.

- Leonard G. Horowitz, author of the self-published works *Emerging Viruses: AIDS & Ebola. Nature, Accident or Intentional?* and *Death in the Air: Globalism, Terrorism and Toxic Warfare*, advances the theory that the AIDS virus was engineered by such U.S. Government defence contractors as Litton Bionetics for the purposes of bio-warfare and "population control."

- Smallpox vaccine theory, In 1987 there was some consideration given to the possibility that the "Aids epidemic may have been triggered by the mass vaccination campaign which eradicated smallpox". An article in the Times suggested this, quoting an unnamed "adviser to WHO" with "*I believe the smallpox vaccine theory is the explanation to the explosion of Aids*". It is now thought that the smallpox vaccine causes serious complications for people who already have impaired immune systems, and the Times article described the case of a military recruit with "dormant HIV" who died within months of receiving it. But no citation was provided regarding people who did not previously have HIV. (HIV is now considered to be a contraindication for the smallpox vaccine- both for an infected person and their sexual partners and household members.) Some conspiracy theorists propose an expanded hypothesis in which the smallpox vaccine was deliberately 'laced' with HIV.

In contrast, a research article was published in 2010 suggesting that it might have been the actual eradication of smallpox and the subsequent *ending* of the mass vaccination campaign that contributed to the sudden emergence of HIV, due to the possibility that immunization against smallpox "might play a role in providing an individual with some degree of protection to subsequent HIV infection and/or disease progression". Regardless of the effects of the smallpox vaccine itself, its use in practice in Africa is one of the categories of un-sterile injection that may have contributed to the spread and mutation of the immunodeficiency viruses.

Alternative Ideas Regarding Causation, Origin or Treatment

- The Duesberg hypothesis promoted by biologist Peter Duesberg argues that AIDS is not caused by HIV, but rather that HIV is a harmless passenger virus, and that AIDS is caused by non-infectious agents like illegal drug usage. The scientific consensus is that the Duesberg hypothesis has been discredited.

- Thabo Mbeki, former President of South Africa, along with other prominent members of the ruling African National Congress party, has argued that AIDS is the result of poverty, chronic disease, malnutrition and other environmental factors. Mbeki based his views on the discredited beliefs of AIDS denialists, especially Peter Duesberg. It has been suggested that the ANC leadership adopted this position as a political expedient, intended to deflect criticism that the ANC had not done enough to fight AIDS in South Africa. In 2000, two statements by government spokespeople, (one later retracted), placed the financial cost of treating pregnant HIV positive women and the subsequent cost to the state of raising the child as central in the decision of whether to provide anti-retroviral drug treatment. Also in 2000, the Johannesburg *Mail &*

Guardian reported that in a leaked text for a speech Mbeki was to give to an ANC caucus, Mbeki claimed that the CIA and Western drug companies were secretly promoting the view that HIV causes AIDS in order to increase sales of anti-AIDS drugs.

Antibiotic Resistance

Antibiotic resistance is a type of drug resistance where a microorganism is able to survive exposure to an antibiotic. Genes can be transferred between bacteria in a horizontal fashion by conjugation, transduction, or transformation. Thus a gene for antibiotic resistance which had evolved via natural selection may be shared. Evolutionary stress such as exposure to antibiotics then selects for the antibiotic resistant trait. Many antibiotic resistance genes reside on plasmids, facilitating their transfer. If a bacterium carries several resistance genes, it is called multiresistant or, informally, a superbug or super bacterium. The primary cause of antibiotic resistance is genetic mutation in bacteria. The prevalence of antibiotic resistant bacteria is a result of antibiotic use both within medicine and veterinary medicine. The greater the duration of exposure the greater the risk of the development of resistance irrespective of the severity of the need for antibiotics. As resistance becomes more common there becomes a greater need for alternative treatments. However despite a push for new antibiotic therapies there has been a continued decline in the number of newly approved drugs. Antibiotic resistance therefore poses a significant problem.

Causes

The widespread use of antibiotics both inside and outside of medicine is playing a significant role in the emergence of resistant bacteria. Antibiotics are often used in rearing animals for food and this use among others leads to the creation of resistant strains of bacteria. In some countries antibiotics are sold over the counter without a prescription which also leads to the creation of resistant strains. In supposedly well-regulated human medicine the major problem of the emergence of resistant bacteria is due to misuse and overuse of antibiotics by doctors as well as patients. Other practices contributing towards resistance include the addition of antibiotics to the feed of livestock. Household use of antibacterials in soaps and other products, although not clearly contributing to resistance, is also discouraged (as not being effective at infection control). Also unsound practices in the pharmaceutical manufacturing industry can contribute towards the likelihood of creating antibiotic resistant strains.

Certain antibiotic classes are highly associated with colonisation with superbugs compared to other antibiotic classes. The risk for colonisation increases if there is a lack of sensitivity (resistance) of the superbugs to the antibiotic used and high tissue penetration as well as broad spectrum activity against "good bacteria". In the case of MRSA, increased rates of MRSA infections are seen with glycopeptides, cephalosporins and especially quinolones. In the case of colonisation with C difficile the high risk antibiotics include cephalosporins and in particular quinolones and clindamycin.

In Medicine

The volume of antibiotic prescribed is the major factor in increasing rates of bacterial resistance rather than compliance with antibiotics. A single dose of antibiotics leads to a greater risk of resistant organisms to that antibiotic in the person for up to a year.

Inappropriate prescribing of antibiotics has been attributed to a number of causes including: people who insist on antibiotics, physicians simply prescribe them as they feel they do not have time to explain why they are not necessary, physicians who do not know when to prescribe antibiotics or else are overly cautious for medical legal reasons. A third of people for example believe that antibiotics are effective for the common cold and 22% of people do not finish a course of antibiotics primarily due to that fact that they feel better (varying from 10% to 44% depending on the country). Compliance with once daily antibiotics is better than with twice daily antibiotics. Sub optimum antibiotic concentrations in critically ill people increase the frequency of antibiotic resistance organisms. While taking antibiotics doses less than those recommended may increase rates of resistance, shortening the course of antibiotics may actually decrease rates of resistance. Poor hand hygiene by hospital staff has been associated with the spread of resistant organisms and an increase in hand washing compliance results in decreased rates of these organisms.

Role of other Animals

Drugs are used in animals that are used as human food, such as cows, pigs, chickens, fish, etc., and these drugs can affect the safety of the meat, milk, and eggs produced from those animals and can be the source of superbugs. For example, farm animals, particularly pigs, are believed to be able to infect people with MRSA. The resistant bacteria in animals due to antibiotic exposure can be transmitted to humans via three pathways, those being through the consumption of meat, from close or direct contact with animals, or through the environment.

The World Health Organization concluded that antibiotics as growth promoters in animal feeds should be prohibited (in the absence of risk assessments). In 1998, European Union health ministers voted to ban four antibiotics widely used to promote animal growth (despite their scientific panel's recommendations). Regulation banning the use of antibiotics in European feed, with the exception of two antibiotics in poultry feeds, became effective in 2006. In Scandinavia, there is evidence that the ban has led to a lower prevalence of antimicrobial resistance in (non-hazardous) animal bacterial populations. In the USA federal agencies do not collect data on antibiotic use in animals but animal to human spread of drug resistant organisms has been demonstrated in research studies. Antibiotics are still used in U.S. animal feed—along with other ingredients which have safety concerns.

Growing U.S. consumer concern about using antibiotics in animal feed has led to a niche market of "antibiotic-free" animal products, but this small market is unlikely to change entrenched industry-wide practices.

In 2001, the Union of Concerned Scientists estimated that greater than 70% of the antibiotics used in the US are given to food animals (e.g. chickens, pigs and cattle) in the absence of disease. In 2000 the US Food and Drug Administration (FDA) announced their intention to revoke approval of fluoroquinolone use in poultry production because of substantial evidence linking it to the emergence of fluoroquinolone resistant campylobacter infections in humans. The final decision to ban fluoroquinolones from use in poultry production was not made until five years later because of challenges from the food animal and pharmaceutical industries. Today, there are two federal bills (S. 549 and H.R. 962) aimed at phasing out "non-therapeutic" antibiotics in US food animal production.

Mechanisms

Schematic representation of how antibiotic resistance evolves via natural selection. The top section represents a population of bacteria before exposure to an antibiotic. The middle section shows the population directly after exposure, the phase in which selection took place. The last section shows the distribution of resistance in a new generation of bacteria. The legend indicates the resistance levels of individuals.

Antibiotic resistance can be a result of horizontal gene transfer, and also of unlinked point mutations in the pathogen genome and a rate of about 1 in 10^8 per chromosomal replication. The antibiotic action against the pathogen can be seen as an environmental pressure; those bacteria which have a mutation allowing them to survive will live on to reproduce. They will then pass this trait to their offspring, which will result in a fully resistant colony.

The four main mechanisms by which microorganisms exhibit resistance to antimicrobials are:

1. Drug inactivation or modification: e.g. enzymatic deactivation of *Penicillin* G in some penicillin-resistant bacteria through the production of β-lactamases.

2. Alteration of target site: e.g. alteration of PBP—the binding target site of penicillins—in MRSA and other penicillin-resistant bacteria.

3. Alteration of metabolic pathway: e.g. some sulfonamide-resistant bacteria do not require para-aminobenzoic acid (PABA), an important precursor for the synthesis of folic acid and nucleic acids in bacteria inhibited by sulfonamides. Instead, like mammalian cells, they turn to utilizing preformed folic acid.

4. Reduced drug accumulation: by decreasing drug permeability and/or increasing active efflux (pumping out) of the drugs across the cell surface.

There are three known mechanisms of fluoroquinolone resistance. Some types of efflux pumps can act to decrease intracellular quinolone concentration. In gram-negative bacteria, plasmid-mediated resistance genes produce proteins that can bind to DNA gyrase, protecting it from the action of quinolones.

Finally, mutations at key sites in DNA gyrase or Topoisomerase IV can decrease their binding affinity to quinolones, decreasing the drug's effectiveness. Research has shown that the bacterial protein LexA may play a key role in the acquisition of bacterial mutations giving resistance to quinolones and rifampicin.

Antibiotic resistance can also be introduced artificially into a microorganism through laboratory protocols, sometimes used as a selectable marker to examine the mechanisms of gene transfer or to identify individuals that absorbed a piece of DNA that included the resistance gene and another gene of interest.

Resistant Pathogens

Staphylococcus Aureus

Staphylococcus aureus (colloquially known as "Staph aureus" or a *Staph infection*) is one of the major resistant pathogens. Found on the mucous membranes and the human skin of around a third of the population, it is extremely adaptable to antibiotic pressure. It was one of the earlier bacteria in which penicillin resistance was found—in 1947, just four years after the drug started being mass-produced. Methicillin was then the antibiotic of choice, but has since been replaced by oxacillin due to significant kidney toxicity. MRSA (methicillin-resistant *Staphylococcus aureus*) was first detected in Britain in 1961 and is now "quite common" in hospitals. MRSA was responsible for 37% of fatal cases of sepsis in the UK in 1999, up from 4% in 1991. Half of all *S. aureus* infections in the US are resistant to penicillin, methicillin, tetracycline and erythromycin.

This left vancomycin as the only effective agent available at the time. However, strains with intermediate (4-8 µg/ml) levels of resistance, termed GISA (glycopeptide intermediate *Staphylococcus aureus*) or VISA (vancomycin intermediate *Staphylococcus aureus*), began appearing in the late 1990s. The first identified case was in Japan in 1996, and strains have since been found in hospitals in England, France and the US. The first documented strain with complete (>16 µg/ml) resistance to vancomycin, termed VRSA (Vancomycin-resistant *Staphylococcus aureus*) appeared in the United States in 2002.

A *new* class of antibiotics, oxazolidinones, became available in the 1990s, and the first commercially available oxazolidinone, linezolid, is comparable to vancomycin in effectiveness against MRSA. Linezolid-resistance in *Staphylococcus aureus* was reported in 2003.

CA-MRSA (Community-acquired MRSA) has now emerged as an epidemic that is responsible for rapidly progressive, fatal diseases including necrotizing pneumonia, severe sepsis and necrotizing fasciitis. Methicillin-resistant *Staphylococcus aureus* (MRSA) is the most frequently identified antimicrobial drug-resistant pathogen in US hospitals. The epidemiology of infections caused by MRSA is rapidly changing. In the past 10 years, infections caused by this organism have emerged in the community. The 2 MRSA clones in the United States most closely associated with community outbreaks, USA400 (MW2 strain, ST1 lineage) and USA300, often contain Panton-Valentine leukocidin (PVL) genes and, more frequently, have been associated with skin and soft tissue infections. Outbreaks of community-associated (CA)-MRSA infections have been reported in correctional facilities, among athletic teams, among military recruits, in newborn nurseries, and among men who have sex with men. CA-MRSA infections now appear to be endemic in many urban regions and cause most CA-S. aureus infections.

Streptococcus and Enterococcus

Streptococcus pyogenes (Group A Streptococcus: GAS) infections can usually be treated with many different antibiotics. Early treatment may reduce the risk of death from invasive group A streptococcal disease. However, even the best medical care does not prevent death in every case. For those with very severe illness, supportive care in an intensive care unit may be needed. For persons with necrotizing fasciitis, surgery often is needed to remove damaged tissue. Strains of *S. pyogenes* resistant to macrolide antibiotics have emerged, however all strains remain uniformly sensitive to penicillin.

Resistance of *Streptococcus pneumoniae* to penicillin and other beta-lactams is increasing worldwide. The major mechanism of resistance involves the introduction of mutations in genes encoding penicillin-binding proteins. Selective pressure is thought to play an important role, and use of beta-lactam antibiotics has been implicated as a risk factor for infection and colonization. Streptococcus pneumoniae is responsible for pneumonia, bacteremia, otitis media, meningitis, sinusitis, peritonitis and arthritis.

Penicillin-resistant pneumonia caused by *Streptococcus pneumoniae* (commonly known as *pneumococcus*), was first detected in 1967, as was penicillin-resistant gonorrhea. Resistance to penicillin substitutes is also known as beyond *S. aureus*. By 1993 *Escherichia coli* was resistant to five fluoroquinolone variants. *Mycobacterium tuberculosis* is commonly resistant to isoniazid and rifampin and sometimes universally resistant to the common treatments. Other pathogens showing some resistance include *Salmonella*, *Campylobacter*, and *Streptococci*.

Enterococcus faecium is another superbug found in hospitals. Penicillin-Resistant Enterococcus was seen in 1983, vancomycin-resistant enterococcus (VRE) in 1987, and Linezolid-Resistant Enterococcus (LRE) in the late 1990s.

Pseudomonas Aeruginosa

Pseudomonas aeruginosa is a highly prevalent opportunistic pathogen. One of the most worrisome characteristics of *P. aeruginosa* consists in its low antibiotic susceptibility. This low susceptibility is attributable to a concerted action of multidrug efflux pumps with chromosomally-encoded antibiotic resistance genes and the low permeability of the bacterial cellular envelopes. Besides intrinsic resistance, *P. aeruginosa* easily develop acquired resistance either by mutation in chromosomally-encoded genes, or by the horizontal gene transfer of antibiotic resistance determinants. Development of multidrug resistance by *P. aeruginosa* isolates requires several different genetic events that include acquisition of different mutations and/or horizontal transfer of antibiotic resistance genes. Hypermutation favours the selection of mutation-driven antibiotic resistance in *P. aeruginosa* strains producing chronic infections, whereas the clustering of several different antibiotic resistance genes in integrons favours the concerted acquisition of antibiotic resistance determinants. Some recent studies have shown that phenotypic resistance associated to biofilm formation or to the emergence of small-colony-variants may be important in the response of *P. aeruginosa* populations to antibiotics treatment.

Clostridium Difficile

Clostridium difficile is a nosocomial pathogen that causes diarrheal disease in hospitals world wide. Clindamycin-resistant *C. difficile* was reported as the causative agent of large outbreaks of diarrheal disease in hospitals in New York, Arizona, Florida and Massachusetts between 1989 and 1992. Geographically dispersed outbreaks of *C. difficile* strains resistant to fluoroquinolone antibiotics, such as Cipro (ciprofloxacin) and Levaquin (levofloxacin), were also reported in North America in 2005.

Salmonella and E. Coli

Escherichia coli and *Salmonella* come directly from contaminated food. Of the meat that is contaminated with *E. coli*, eighty percent of the bacteria are resistant to one or more drugs made; it causes bladder infections that are resistant to antibiotics ("HSUS Fact Sheet"). *Salmonella* was first found in humans in the 1970s and in some cases is resistant to as many as nine different antibiotics ("HSUS Fact Sheet"). When both bacterium are spread, serious health conditions arise. Many people are hospitalized each year after becoming infected, and some die as a result.

Acinetobacter Baumannii

On November 5, 2004, the Centres for Disease Control and Prevention (CDC) reported an increasing number of *Acinetobacter baumannii* bloodstream infections in patients at military medical facilities in which service members injured in the Iraq/Kuwait region during Operation Iraqi Freedom and in Afghanistan during Operation Enduring Freedom were treated. Most of these showed multidrug resistance (MRAB), with a few isolates resistant to all drugs tested.

Alternatives

Prevention

Rational use of antibiotics may reduce the chances of development of opportunistic infection by antibiotic-resistant bacteria due to dysbacteriosis. In one study the use of fluoroquinolones are clearly associated with *Clostridium difficile* infection, which is a leading cause of nosocomial diarrhea in the United States, and a major cause of death, worldwide.

There is clinical evidence that topical dermatological preparations such as those containing tea tree oil and thyme oil may be effective in preventing transmittal of CA-MRSA. In addition, other phytotherapeutic medicines too can reduce the use of antibiotics or eliminate their use entirely.

Vaccines do not suffer the problem of resistance because a vaccine enhances the body's natural defences, while an antibiotic operates separately from the body's normal defences. Nevertheless, new strains may evolve that escape immunity induced by vaccines; for example an update Influenza vaccine is needed each year.

While theoretically promising, anti-staphylococcal vaccines have shown limited efficacy, because of immunological variation between *Staphylococcus* species, and the limited duration of effectiveness of the antibodies produced. Development and testing of more effective vaccines is under way.

The Australian Commonwealth Scientific and Industrial Research Organization (CSIRO), realizing the need for the reduction of antibiotic use, has been working on two alternatives. One alternative is to prevent diseases by adding cytokines instead of antibiotics to animal feed. These proteins are made in the animal body "naturally" after a disease and are not antibiotics so they do not contribute to the antibiotic resistance problem. Furthermore, studies on using cytokines have shown that they also enhance the growth of animals like the antibiotics now used, but without the drawbacks of non-therapeutic antibiotic use. Cytokines have the potential to achieve the animal growth rates traditionally sought by the use of antibiotics without the contribution of antibiotic resistance associated with the widespread non-therapeutic uses of antibiotics currently utilized in the food animal production industries. Additionally, CSIRO is working on vaccines for diseases.

Phage Therapy

Phage therapy, an approach that has been extensively researched and utilized as a therapeutic agent for over 60 years, especially in the Soviet Union, is an alternative that might help with the problem of resistance. Phage therapy was widely used in the United States until the discovery of antibiotics, in the early 1940s. Bacteriophages or "phages" are viruses that invade bacterial cells and, in the case of lytic phages, disrupt bacterial metabolism and cause the bacterium to lyse. Phage therapy is the therapeutic use of lytic bacteriophages to treat pathogenic bacterial infections.

Bacteriophage therapy is an important alternative to antibiotics in the current era of multidrug resistant pathogens. A review of studies that dealt with the therapeutic use of phages from 1966–1996 and few latest ongoing phage therapy projects via internet showed: phages were used topically, orally or systemically in Polish and Soviet studies. The success rate found in these studies was 80–95% with few gastrointestinal or allergic side effects. British studies also demonstrated significant efficacy of phages against *Escherichia coli*, *Acinetobacter* spp., *Pseudomonas* spp and *Staphylococcus aureus*. US studies dealt with improving the bioavailability of phage. Phage therapy may prove as an important alternative to antibiotics for treating multidrug resistant pathogens.

Research

New Medications

Until recently, research and development (R&D) efforts have provided new drugs in time to treat bacteria that became resistant to older antibiotics. That is no longer the case. The potential crisis at hand is the result of a marked decrease in industry R&D, and the increasing prevalence of resistant bacteria. Infectious disease physicians are alarmed by the prospect that effective antibiotics may not be available to treat seriously ill patients in the near future.

The pipeline of new antibiotics is drying up.Major pharmaceutical companies are losing interest in the antibiotics market because these drugs may not be as profitable as drugs that treat chronic (long-term) conditions and lifestyle issues.

The resistance problem demands that a renewed effort be made to seek antibacterial agents effective against pathogenic bacteria resistant to current antibiotics. One of the possible strategies towards this objective is the rational localization of bioactive phytochemicals. Plants have an almost limitless ability to synthesize aromatic substances, most of which are phenols or their oxygen-substituted derivatives such as tannins. Most are secondary metabolites, of which at least 12,000 have been isolated, a number estimated to be less than 10% of the total. In many cases, these substances serve as plant defence mechanisms against predation by microorganisms, insects, and herbivores. Many of the herbs and spices used by humans to season food yield useful medicinal compounds including those having antibacterial activity.

Traditional healers have long used plants to prevent or cure infectious conditions. Many of these plants have been investigated scientifically for antimicrobial activity and a large number of plant products have been shown to inhibit growth of pathogenic bacteria.A number of these agents appear to have structures and modes of action that are distinct from those of the antibiotics in current use, suggesting that cross-resistance with agents already in use may be minimal. For example the combination of 5'-methoxyhydnocarpine and berberine in herbs like Hydrastis canadensis and Berberis vulgaris can block the MDR-pumps that cause multidrug resistance. This has been shown for Staphylococcus aureus.

Archaeocins is the name given to a new class of potentially useful antibiotics that are derived from the Archaea group of organisms. Eight archaeocins have been partially or fully characterized, but hundreds of archaeocins are believed to exist, especially within the haloarchaea. The prevalence of archaeocins is unknown simply because no one has looked for them. The discovery of new archaeocins hinges on recovery and cultivation of archaeal organisms from the environment. For example, samples from a novel hypersaline field site, Wilson Hot Springs, recovered 350 halophilic organisms; preliminary analysis of 75 isolates showed that 48 were archaeal and 27 were bacterial.

In research published on October 17, 2008 in *Cell*, a team of scientists pinpointed the place on bacteria where the antibiotic myxopyronin launches its attack, and why that attack is successful. The myxopyronin binds to and inhibits the crucial bacterial enzyme, RNA polymerase. The myxopyronin changes the structure of the switch-2 segment of the enzyme, inhibiting its function of reading and transmitting DNA code.

This prevents RNA polymerase from delivering genetic information to the ribosomes, causing the bacteria to die. One of the major causes of antibiotic resistance is the decrease of effective drug concentrations at their target place, due to the increased action of ABC transporters. Since ABC transporter blockers can be used in combination with current drugs to increase their effective intracellular concentration, the possible impact of ABC transporter inhibitors is of great clinical interest. ABC transporter blockers that may be useful to increase the efficacy of current drugs have entered clinical trials and are available to be used in therapeutic regimes.

Applications

Antibiotic resistance is an important tool for genetic engineering. By constructing a plasmid which contains an antibiotic resistance gene as well as the gene being engineered or expressed, a researcher can ensure that when bacteria replicate, only the copies which carry along the plasmid survive. This ensures that the gene being manipulated passes along when the bacteria replicates. The most commonly used antibiotics in genetic engineering are generally "older" antibiotics which have largely fallen out of use in clinical practice. These include:

- ampicillin
- kanamycin
- tetracycline
- chloramphenicol.

Industrially the use of antibiotic resistance is disfavored since maintaining bacterial cultures would require feeding them large quantities of antibiotics. Instead, the use of auxotrophic bacterial strains (and function-replacement plasmids) is preferred.

Biological Hazard

Biological hazards, also known as biohazards, refer to biological substances that pose a threat to the health of living organisms, primarily that of humans. This can include medical waste or samples of a microorganism, virus or toxin (from a biological source) that can impact human health. It can also include substances harmful to animals. The term and its associated symbol is generally used as a warning, so that those potentially exposed to the substances will know to take precautions.

In Unicode, the biohazard sign is U+2623.

Biohazardous agents are classified for transportation by UN number:

- UN 2814 (Infectious substance to Humans)
- UN 2900 (Infectious substance to Animals)
- UN 3291 (Medical Waste)
- Category A, UN 2814-Infectious substances affecting humans and animals: An infectious substance in a form capable of causing permanent disability or life-threatening or fatal disease in otherwise healthy humans or animals when exposure to it occurs.
- Category B, UN 2900-Infectious substances affecting animals only: An infectious substance that is not in a form generally capable of causing permanent disability of life-threatening or fatal disease in otherwise healthy humans and animals when exposure to themselves occurs.
- Category B, UN 3373-Biological substance transported for diagnostic or investigative purposes.
- Regulated Medical Waste, UN 3291-Waste or reusable material derived from medical treatment of an animal or human, or from biomedical research, which includes the production and testing of biological products.

Levels of Biohazard

The United States' Centres for Disease Control and Prevention (CDC) categorizes various diseases in levels of biohazard, Level 1 being minimum risk and Level 4 being extreme risk. Laboratories and other facilities are categorized as BSL (Biosafety Level) 1-4 or as *P1* through *P4* for short (Pathogen or Protection Level).

- Biohazard Level 1: Bacteria and viruses including *Bacillus subtilis*, canine hepatitis, *Escherichia coli*, varicella (chicken pox), as well as some cell cultures and non-infectious bacteria. At this level precautions against the biohazardous materials in question are minimal, most likely involving gloves and some sort of facial protection. Usually, contaminated materials are left in open (but separately indicated) waste receptacles. Decontamination procedures for this level are similar in most respects to modern precautions against everyday viruses (i.e.: washing one's hands with anti-bacterial soap, washing all exposed surfaces of the lab with disinfectants, etc). In a lab environment, all materials used for cell and/or bacteria cultures are decontaminated via autoclave.

- Biohazard Level 2: Bacteria and viruses that cause only mild disease to humans, or are difficult to contract via aerosol in a lab setting, such as hepatitis A, B, and C, influenza A,Lyme disease, salmonella, mumps, measles, scrapie, dengue fever, and HIV. "Routine diagnostic work with clinical specimens can be done safely at Biosafety Level 2, using Biosafety Level 2 practices and procedures. Research work (including co-cultivation, virus replication studies, or manipulations involving concentrated virus) can be done in a BSL-2 (P2) facility, using BSL-3 practices and procedures. Virus production activities, including virus concentrations, require a BSL-3 (P3) facility and use of BSL-3 practices and procedures".

- Biohazard Level 3: Bacteria and viruses that can cause severe to fatal disease in humans, but for which vaccines or other treatments exist, such as anthrax, West Nile virus, Venezuelan equine encephalitis, SARS virus, variola virus (smallpox), tuberculosis, typhus, Rift Valley fever, Rocky Mountain spotted fever, yellow fever, and malaria. Among parasites *Plasmodium falciparum*, which causes Malaria, and *Trypanosoma cruzi*, which causes trypanosomiasis, also come under this level.

- Biohazard Level 4: Viruses and bacteria that cause severe to fatal disease in humans, and for which vaccines or other treatments are *not* available, such as Bolivian and Argentine hemorrhagic fevers, (Argentine Hemorragic does have a vaccine "The Candid #1 vaccine for AHF was created in 1985 by Argentine virologist Dr. Julio Barrera Oro. The vaccine was manufactured by the Salk Institute in the United States, and became available in Argentina since 1990.

Candid #1 has been applied to adult high-risk population and is 95.5% effective. On 29 August 2006, the Maiztegui Institute obtained certification for the production of the vaccine in Argentina. A vaccination plan is yet to be outlined, but the budget for 2007 allows for 390,000 doses, at AR$8 each (about US$2.6 or €2). The Institute has the capacity to manufacture, in one year, the 5 million doses required to vaccinate the entire population of the endemic area." Argentine_hemorrhagic_fever#Vaccine H5N1(bird flu), Dengue hemorrhagic fever, Marburg virus, Ebola virus, hantaviruses, Lassa fever, Crimean-Congo hemorrhagic fever, and other hemorrhagic diseases. When dealing with biological hazards at this level the use of a Hazmat suit and a self-contained oxygen supply is mandatory. The entrance and exit of a Level Four biolab will contain multiple showers, a vacuum room, an ultraviolet light room, autonomous detection system, and other safety precautions designed to destroy all traces of the biohazard. Multiple airlocks are employed and are electronically secured to prevent both doors opening at the same time. All air and water service going to and coming from a Biosafety Level 4 (P4) lab will undergo similar decontamination procedures to eliminate the possibility of an accidental release.

Biosecurity

Biosecurity is a set of preventive measures designed to reduce the risk of transmission of infectious diseases, quarantined pests, invasive alien species, living modified organisms. While biosecurity does encompass the prevention of the intentional removal (theft) of biological materials from research laboratories, this definition is narrower in scope than the definition used by many experts, including the United Nations Food and Agriculture Organization. These preventative measures are a combination of systems and practices put into its place at legitimate bioscience laboratories to prevent the use of dangerous pathogens and toxins for malicious use, as well as by customs agents and agricultural and natural resource managers to prevent the spread of these biological agents in natural and managed. Reference no. 123 ecosystems. Although security is usually thought of in terms of "Guards, Gates, and Guns", biosecurity encompasses much more than that and requires the cooperation of scientists, technicians, policy makers, security engineers, and law enforcement officials.

Components of a laboratory biosecurity program include:

- Physical security

- Personnel security
- Material control & accountability
- Transport security
- Information security
- Program management.

Animal Biosecurity

Animal biosecurity is the product of all actions undertaken by an entity to prevent introduction of disease agents into a specific area. Animal biosecurity differs from biosecurity which are measures taken to reduce the risk of infectious agent theft and dispersal by means of bioterrorism. Animal biosecurity is a comprehensive approach, encompassing different means of prevention and containment. A critical element in animal biosecurity, biocontainment, is the control of disease agents already present in a particular area, and works to prevent novel transmissions. Animal biosecurity may protect organisms from infectious agents or noninfectious agents such as toxins or pollutants, and can be executed in areas as large as a nation or as small as a local farm.

Animal biosecurity takes into account the epidemiological triad for disease occurrence: the individual host, the disease, and the environment in contributing to disease susceptibility. It aims to improve nonspecific immunity of the host to resist the introduction of an agent, or limit the risk that an agent will be sustained in an environment at adequate levels. Biocontainment, an element of animal biosecurity, works to improve specific immunity towards already present pathogens.

Biosecurity means the prevention of the illicit use of pathogenic bioorganisms by laboratory staff or others. Biosafety means the protection of laboratory staff from being infected by pathogenic bioorganisms.

Challenges

The destruction of the World Trade Centre in Manhattan on September 11, 2001 by terrorists, and subsequent wave of anthrax attacks on U.S. media and government outlets (both real and hoax), led to increased attention on the risk of bioterror attacks in the United States. Proposals for serious structural reforms, national and/or regional border controls, and a single co-ordinated system of biohazard response abounded.

One of the major challenges in biosecurity is the increasing availability and accessibility of potentially harmful technology. Biomedical advances and the globalization of scientific and technical expertise have made it possible to greatly improve public health. However, there is also the risk that advances can lead to make biological weapons.

The proliferation of high biosafety level laboratories around the world has many experts worried about availability of targets for those that might interested in stealing dangerous pathogens. Emerging and Re-emerging disease is also a serious biosecurity concern. The recent growth in containment laboratories is often in response to emerging diseases, many new containment lab's main focus is to find ways to control these diseases. By strengthening national disease surveillance, prevention, control and response systems, these labs are raising international public health to new heights.

UNU/IAS Research into Biosecurity & Biosafety emphasizes "long-term consequences of the development and use of biotechnology" and need for "an honest broker to create avenues and forums to unlock the impasses."

Biosecurity Incidents

- 1984 Rajneeshee religious cult attacks, The Dalles, Oregon
- Objective: Gain control of the Wasco County Court by affecting the election
- Organism: Salmonella typhimurium, purchased from commercial supplier
- Dissemination: Restaurant salad bars
- 751 illnesses, Early investigation by CDC suggested the event was a naturally occurring outbreak. Cult member arrested on unrelated charge confessed involvement with the event.
- 1990s Aum Shinrikyo attempts in Tokyo, Japan:
- Objective: Fulfill apocalyptic prophecy
- Organisms: Bacillus anthracis (Vaccine strain), Clostridium botulinum (Environmental isolate, Avirulent strain) Aum Shinrikyo ordered Clostridium botulinum from a pharmaceutical company, and Ebola virus (Attempted to acquire from Zaire outbreak under guise of an "Humanitarian mission")
- Dissemination: Aerosolization in Tokyo (B. anthracis and Botulinum toxin)

- Leader Asahara was convicted of criminal activity.
  o 2001 Anthrax attacks in the US
  o 1995—Larry Wayne Harris, a white supremacist, ordered 3 vials of Yersinia pestis from the ATCC
  o 1995—Laboratory technician Diane Thompson removed Shigella dysenteriae Type 2 from hospital's collection and infected co-workers
  o Professor Thomas Butler, United States, 2003.
    * 30 vials of Yersinia pestis missing from lab (never recovered); Butler served 19 months in jail.
    * Dr. Mario Jascalevich, New Jersey doctor, accused of poisoning 5 patients with this plant-derived toxin-Tubocurarine: 1966
    * Arnfinn Nesset, nursing home operator in Norway, killed 27 residents at a nursing home with curacit: May 1977 – November 1980
    * Dr. David Acer, Florida dentist, infects 6 patients with HIV, unclear if this was a deliberate act: 1987–1990
    * Dr. Ray W. Mettetal, Jr., a neurologist in Virginia, was found in possession of ricin after arrest on another issue: 1995
    * Debora Green, a physician, convicted of trying to murder her estranged husband with ricin
    * Richard Schmidt, a gastroenterologist in Louisiana, convicted of attempted second degree murder for infecting nurse Janice Allen with HIV by injecting her with blood from an AIDS patient: 1998
    * Brian T. Stewart, a phlebotomist, sentenced to life in prison for deliberately infecting his 11-month-old baby with HIV-infected blood to avoid child support payments: 1999
    * Physician reports theft of a vial of Mycobacterium tuberculosis: June 1999
    * Japan 1964-1966, Dr. Mitsuru Suzuki was a physician with training in bacteriology
    * Objective: Revenge due to deep antagonism to what he perceived as a prevailing seniority system

* Organisms: Shigella dysenteriae and Salmonella typhi
* Dissemination: Sponge cake, other food sources
* He was later implicated in 200 – 400 illnesses and 4 deaths
* Official investigation started after anonymous tip to Ministry of Health and Welfare. He was charged, but was not convicted of any deaths

- Hospital in Dallas, TX, 1996, Diane Thompson was a clinical laboratory technician
  o Objective: Unclear, possibly revenge against former boyfriend and cover-up by infecting co-workers
  o Organism: Shigella dysenteriae Type 2, acquired from clinical laboratory
  o Dissemination: Contaminated pastries in the office breakroom
  o Infected 12 of her coworkers, She was arrested, convicted, 20 year sentence.

The Role of Education in Biosecurity

The advance of the life sciences and biotechnology has the potential to bring great benefits to humankind through responding to societal challenges. However, it is also possible that such advances could be exploited for hostile purposes, something evidenced in a small number of incidents of bioterrorism, but more particularly by the series of large-scale offensive biological warfare programmes carried out by major states in the last century.

Dealing with this challenge, which has been labelled the 'dual-use' dilemma requires a number of different activities such as those identified above as being require for biosecurity. However, one of the essential ingredients in ensuring that the life sciences continue to generate great benefits and do not become subject to misuse for hostile purposes is a process of engagement between scientists and the security community and the development of strong ethical and normative frameworks to compliment legal and regulatory measures that are being developed by states.

Biosecurity Regulations

- US Select Agent Regulations:
  o Facility registration if it possesses one of 81 Select Agents

- o Facility must designate a Responsible Official
- o Background checks for individuals with access to Select Agents
- o Access controls for areas and containers that contain Select Agents
- o Detailed inventory requirements for Select Agents
- o Security, safety, and emergency response plans
- o Safety and security training
- o Regulation of transfers of Select Agents
- o Extensive documentation and recordkeeping
- o Safety and security inspections.
- Biological Weapons Convention addresses three relevant issues:
  - o National Implementing Legislation
  - o National Pathogen Security (biosecurity)
  - o International Cooperation
  - o States Parties agree to pursue national implementation of laboratory and transportation biosecurity (2003)
- UN 1540:
  - o Urges States to take preventative measures to mitigate the threat of WMD proliferation by non-state actors
  - o "Take and enforce effective measures to establish domestic controls to prevent the proliferation of... biological weapons...; including by establishing appropriate controls over related materials"
  - o European Commission Green Paper on Bio-Preparedness (November 2007)
  - o recommends developing European standards on laboratory biosecurity including Physical protection, access controls, accountability of pathogens, and registration of researchers.
- Organization for Economic Cooperation and Development:
  - o published "Best Practice Guidelines for Biological Resource Centres" including a section on biosecurity in February 2007
- Kampala Compact (October 2005) and the Nairobi Announcement (July 2007)
  - o Stress importance of implementing laboratory biosafety and biosecurity in Africa.

Agents of Concern

The following agents are deemed a biosecurity concern by the US Government through the US Select Agent List. The list is divided between agents that can infect only humans, zoonotic agents (which can infect both humans and animals), agents that can only infect animals, and agents that can infect only plants.

Human Agents

- Abrin
- Cercopithecine herpesvirus 1 (Herpes B Virus)
- Coccidioides posadasii
- Conotoxins
- Crimean-Congo haemorrhagic fever virus
- Diacetoxyscirpenol
- Ebola virus
- Lassa fever virus
- Marburg virus
- Monkeypox virus
- Reconstructed replication competent forms of the 1918 pandemic influenza virus containing any portion of the coding regions of all eight gene segments (Reconstructed 1918 Influenza virus)
- Ricin
- Rickettsia prowazekii
- Rickettsia rickettsii
- Saxitoxin
- Shiga-like ribosome inactivating proteins
- South American Haemorrhagic Fever viruses;
  - o Flexal Virus
  - o Guanarito virus
  - o Junin virus
  - o Machupo
  - o Sabia
- Tetrodotoxin
- Tick-borne encephalitis complex (flavi) viruses
  - o Central European Tick-borne encephalitis

- o Far Eastern Tick-borne encephalitis
- o Kyasanur Forest disease
- o Omsk hemorrhagic fever
- o Russian spring-summer encephalitis
- o Variola major virus (Smallpox virus) and
- o Variola minor virus (Alastrim)
- o Yersinia pestis.

Zoonotic Agents

(Overlap Select Agents And Toxins):

- Bacillus anthracis
- Botulinum neurotoxin
- Botulinum neurotoxin producing species of Clostridium
- Brucella abortus
- Brucella melitensis
- Brucella suis
- Burkholderia mallei (formerly Pseudomonas mallei)
- Burkholderia pseudomallei (formerly Pseudomonas pseudomallei)
- Clostridium perfringens epsilon toxin
- Coccidioides immitis
- Coxiella burnetii
- Eastern equine encephalitis virus
- Francisella tularensis
- Hendra virus
- Nipah virus
- Rift Valley fever virus
- Shiga toxin
- Staphylococcal enterotoxins
- T-2 toxin
- Venezuelan equine encephalitis virus.

Animal agents

(Usda select agents and Toxins):

- African horse sickness virus

- African swine fever virus
- Akabane virus
- Avian influenza virus (highly pathogenic)
- Bluetongue virus (Exotic)
- Bovine spongiform encephalopathy
- Camel pox virus
- Classical swine fever virus
- Cowdria ruminantium (Heartwater)
- Foot-and-mouth disease virus
- Goat pox virus
- Japanese encephalitis virus
- Lumpy skin disease virus
- Malignant catarrhal fever virus (Alcelaphine herpesvirus type 1)
- Menangle virus
- Mycoplasma capricolum/M.F38/M. mycoides Capri (contagious caprine pleuropneumonia)
- Mycoplasma mycoides mycoides (contagious bovine pleuropneumonia)
- Newcastle disease virus (velogenic)
- Peste des petits ruminants virus
- Rinderpest virus
- Sheep pox virus
- Swine vesicular disease virus
- Vesicular stomatitis virus (Exotic).

Plant Agents

(Usda plant protection and quarantine (PPQ) select agents and toxins):

- Candidatus *Liberobacter africanus*
- Candidatus *Liberobacter asiaticus*
- *Peronosclerospora philippinensis*
- *Ralstonia solanacearum* race 3, biovar. 2
- *Sclerophthora rayssiae* var. *zeae*
- *Synchytrium endobioticum*

- *Xanthomonas oryzae* pv. *oryzicola*
- *Xylella fastidiosa* (citrus variegated chlorosis strain).

Entomological Warfare

Entomological warfare (EW) is a type of biological warfare that uses insects to attack the enemy. The concept has existed for centuries and research and development have continued into the modern era. EW has been used in battle by Japan and several other nations have developed and been accused of using an entomological warfare program.

Description

Entomological warfare (EW) is a specific type of biological warfare (BW) that uses insects in a direct attack or as vectors to deliver a biological agent, such as plague or cholera. Essentially, EW exists in three varieties. One type of EW involves infecting insects with a pathogen and then dispersing the insects over target areas.

The insects then act as a vector, infecting any person or animal they might bite. Another type of EW is a direct insect attack against crops; the insect may not be infected with any pathogen but instead represents a threat to agriculture. The final method of entomological warfare is to use uninfected insects, such as bees, to directly attack the enemy.

History

Entomological warfare is not a new concept; historians and writers have studied EW in connection to multiple historic events. A 14th century plague epidemic in Asia Minor that eventually became known as the Black Death is one such event that has drawn attention from historians as a possible early incident of entomological warfare. That plague's spread over Europe may have been the result of a biological attack on the Crimean city of Kaffa. According to Jeffrey Lockwood, author of *Six-Legged Soldiers* (a book about EW), the earliest incident of entomological warfare was probably the use of bees by early humans.

The bees or their nests were thrown into caves to force the enemy out and into the open. Lockwood theorizes that the Ark of Covenant may have been deadly when opened because it contained deadly fleas.

During the American Civil War the Confederacy accused the Union of purposely introducing the harlequin bug in the South. These accusations were never proven, and modern research has shown it more likely that the insect arrived by other means. The world did not experience large-scale entomological warfare until World War II; Japanese attacks in China were the only verified instance of BW or EW during the war. During, and following, the war other nations began their own EW programs.

World War II

Canada

Among the Allied Powers, Canada led the pioneering effort in vector-borne warfare. After Japan became intent on developing the plague flea as a weapon, Canada, and then the United States followed suit. Cooperating closely with the United States, Dr. G.B. Reed, chief of Kingston's Queen's University's Defence Research Laboratory, focused his research efforts on mosquito vectors, biting flies, and plague infected fleas during World War II. Much of this research was shared with or conducted in concert with the United States.

Canada's entire bio-weapons program was ahead of the British and the Americans during the war. The Canadians tended to work in areas their allies ignored, entomological warfare was one of these areas. As the U.S. and British programs evolved, the Canadians worked closely with both nations. The Canadian BW work would continue well after the war, including entomological research.

France

France is known to have pursued entomological warfare programs during World War II. Like Germany, the nation suggested that the Colorado potato beetle, aimed at the enemy's food sources, would be an asset during the war. As early as 1939 biological warfare experts in France suggested that the beetle be used against German crops.

Germany

Germany is known to have pursued entomological warfare programs during World War II. The nation pursued the mass-production, and dispersion, of the Colorado potato beetle (*Lepinotarsa decemlineata*), aimed at the enemy's food sources. The beetle was first found in Germany in 1914, as an invasive species from North America. There are no records that indicate the beetle was ever employed as a weapon by Germany, or any other nation during the war. Regardless, the Germans had developed plans to drop the beetles on English crops.

Germany carried out testing of its Colorado potato beetle weaponization program south of Frankfurt, where they released 54,000 of the beetles. In 1944, an infestation of Colorado potato beetles was reported in Germany. The source of the infestation is unknown, speculation has offered three alternative theories as to the origin of the infestation. One option is Allied action, an entomological attack, another is that it was the result of the German testing, and still another more likely explanation is that it was merely a natural occurrence.

Japan

Japan used entomological warfare on a large-scale during World War II in China. Unit 731, Japan's infamous biological warfare unit, used plague-infected fleas and flies covered with cholera to infect the population in China. The Japanese military dispersed the insects by spraying them from low-flying airplanes and dropping bombs filled with a mixture of insects and disease. Localized and deadly epidemics resulted and nearly 500,000 Chinese died of disease. An "international symposium" of historians declared in 2002 that Japanese entomological warfare in China was responsible for the deaths of 440,000.

United Kingdom

A British scientist, J.B.S. Haldane, suggested that Britain and Germany were both vulnerable to entomological attack via the Colorado potato beetle. In 1942 the United States shipped 15,000 Colorado potato beetles to Britain for study as a weapon.

Cold War

Soviet Union

The Soviet Union researched, developed and tested an entomological warfare program as a major part of an anti-crop and anti-animal BW program. The Soviets developed techniques for using insects to transmit animal pathogens, such as foot and mouth disease — which they used ticks to transmit. The nation also used avian ticks to transmit *Chlamydophila psittaci* to chickens. In addition, the Soviet Union claimed to have developed an automated mass insect breeding facility, capable of outputting millions of parasitic insects per day.

United States

The United States seriously researched the potential of entomological warfare during the Cold War. The United States military developed plans for an entomological warfare facility, designed to produce 100 million yellow fever-infected mosquitoes per month. A

U.S. Army report titled "Entomological Warfare Target Analysis" listed vulnerable sites within the Soviet Union that the U.S. could attack using entomological vectors. The military also tested the mosquito biting capacity by dropping uninfected mosquitoes over U.S. cities.

North Korean and Chinese officials levelled accusations that during the Korean War the United States engaged in biological warfare, including EW, in North Korea. The claim is dated to the period of the war, and has been thoroughly denied by the U.S. In 1998, Stephen Endicott and Edward Hagermann claimed that the accusations were true in their book, *The United States and Biological Warfare: Secrets from the Early Cold War and Korea* The book received mixed reviews, some called it "bad history" and "appalling", while other praised the case the authors made. Other historians have revived the claim in recent decades as well. The same year Endicotts' book was published Kathryn Weathersby and Milton Leitenberg of the Cold War International History Project at the Woodrow Wilson Centre in Washington released a cache of Soviet and Chinese documents which revealed the North Korean claim was an elaborate disinformation campaign.

During the 1950s the United States conducted a series of field tests using entomological weapons. Operation Big Itch, in 1954, was designed to test munitions loaded with uninfected fleas (*Xenopsylla cheopis*). Big Itch went awry when some of the fleas escaped into the plane and bit all three members of the air crew. In May 1955 over 300,000 yellow fever mosquitoes (*Aedes aegypti*) were dropped over parts of the U.S. state of Georgia to determine if the air-dropped mosquitoes could survive to take meals from humans. The mosquito tests were known as Operation Big Buzz. The U.S. engaged in at least two other EW testing programs, Operation Drop Kick and Operation May Day. A 1981 Army report outlined these tests as well as multiple cost-associated issues that occurred with EW. The report is partially declassified — some information is blacked out, including everything concerning "Drop Kick" — and included "cost per death" calculations. The cost per death, according to the report, for a vector-borne biological agent achieving a 50% mortality rate in an attack on a city was $0.29 in 1976 dollars. Such an attack was estimated to result in 625,000 deaths.

The United States has also applied entomological warfare research and tactics in non-combat situations. In 1990 the U.S. funded a $6.5

million program designed to research, breed and drop caterpillars. The caterpillars were to be dropped in Peru on coca fields as part of the American War on Drugs. As recently as 2002 U.S. entomological anti-drug efforts at Fort Detrick were focused on finding an insect vector for a virus that affects the opium poppy.

Bioterrorism

Clemson University's Regulatory and Public Service Program listed "diseases vectored by insects" among bioterrorism scenarios considered "most likely". Because invasive species are already a problem worldwide one University of Nebraska entomologist considered it likely that the source of any sudden appearance of a new agricultural pest would be difficult, if not impossible, to determine. Lockwood considers insects a more effective means of transmitting biological agents for acts of bioterrorism than the actual agents. Insect vectors are easily gathered and their eggs easily transportable without detection. Isolating and delivering biological agents, on the other hand, is extremely challenging and hazardous.

In one of the few suspected acts of entomological bioterrorism an eco-terror group known as The Breeders claimed to have released Mediterranean fruit flies (medflies) amidst an ongoing California infestation. Lockwood asserts that there is some evidence the group played a role in the event.

The pest attacks a variety of crops and the state of California responded with a large-scale pesticide spraying program. At least one source asserted that there is no doubt that an outside hand played a role in the dense 1989 infestation. The group stated in a letter to then Los Angeles Mayor Tom Bradley that their goals were twofold. They sought to cause the medfly infestation to grow out of control which, in turn, would render the ongoing Malathion spraying program financially infeasible.

Legal Status

The Biological and Toxic Weapons Convention (BWC) of 1972 does not specifically mention insect vectors in its text. The language of the treaty, however, does cover vectors. Article I bans, "Weapons, equipment or means of delivery designed to use such agents or toxins for hostile purposes or in armed conflict." It would appear, due to the text of the BWC, that insect vectors as an aspect of entomological warfare are covered and outlawed by the Convention. The issue is less clear when warfare with uninfected insects against crops is considered.

Ethnic Bioweapon

An ethnic bioweapon (*biogenetic weapon*) aims to harm only or primarily persons of specific ethnicities or genotypes.

History of Ethnic Bioweapons

One of the first fictional discussions of ethnic weapons is in Robert A. Heinlein's 1942 novel *Sixth Column* (republished as *The Day After Tomorrow*) in which a race-specific radiation weapon is used against a so-called "Pan-Asian" invader.

Genetic Weapon

In 1997, U.S. Secretary of Defence William Cohen referred to the concept as a possible risk. In 1998 some biological weapon experts considered such a "genetic weapon" a plausible possibility, and believed the former Soviet Union had undertaken some research on the influence of various substances on human genes.

The possibility of a "genetic bomb" is presented in Vincent Sarich's and Frank Miele's book, *Race: The Reality of Human Differences*, published in 2004. The authors believe that information from the Human Genome Project will be used in just such a manner.

In 2005 the official view of the International Committee of the Red Cross was "The potential to target a particular ethnic group with a biological agent is probably not far off. These scenarios are not the product of the ICRC's imagination but have either occurred or been identified by countless independent and governmental experts."

Israeli "Ethno-bomb" Controversy

In November 1998, *The Sunday Times* reported that Israel was attempting to build an "ethno-bomb" containing a biological agent that could specifically target genetic traits present amongst Arab populations. *Wired News* also reported the story, as did *Foreign Report*. The article was quickly denounced as a hoax. Microbiologists and geneticists were skeptical towards the scientific plausibility of such a biological agent. The *New York Post*, describing the claims as "blood libel", reported that the likely source for the story was a work of science fiction by Israeli academic Doron Stanitsky. Stanitsky had sent his completely fictional work about such a weapon to Israeli newspapers two years before. The article also noted the views of genetic researchers who claimed the idea as "wholly fantastical", still others admit that the weapon was theoretically possible. A planned second installment of the article never appeared, and no sources were ever identified. Neither of the authors of the Sunday Times story, Uzi Mahnaimi and Marie Colvin, have spoken publicly on the matter.

Russian Ban on Export of Biological Samples

In May 2007, Russian newspaper *Kommersant* reported that the Russian government banned all exports of human biosamples. The report claims that the reason for the ban was a secret FSB report about on-going development of "genetic bioweapons" targeting Russian population by Western institutions. The report mentions the Harvard School of Public Health, American International Health Alliance, United States Department of Justice Environment and Natural Resources Division, Karolinska Institutet and United States Agency for International Development.

Deinococcus Radiodurans: A Marvelous Berry that Withstands Radiations

Deinococcus radiodurans is an extremophilic bacterium, one of the most radioresistant organisms known. It can survive cold, dehydration, vacuum, and acid, and is therefore known as a polyextremophile and has been listed as the world's toughest bacterium in *The Guinness Book Of World Records*.

Name and Classification

The name *Deinococcus radiodurans* derives from the Greek deinos and kokkos meaning "terrible grain" and the Latin radius and durare meaning "radiation surviving". The species was formerly called *Micrococcus radiodurans*. As a consequence of its hardiness, it has been nicknamed "Conan the Bacterium", a play on "Conan the Barbarian".

Initially it was placed in the genus *Micrococcus*. After evaluation of ribosomal RNA sequences and other evidence, it was placed in its own genus *Deinococcus*, which is closely related to the genus *Thermus* of heat-resistant bacteria; the group consisting of the two is accordingly known as Deinococcus-Thermus.

Deinococcus is the only genus in the order *Deinococcales*. *D. radiodurans* is the type species of this genus, and the best studied member. All known members of the genus are radioresistant: *D. proteolyticus, D. radiopugnans, D. radiophilus, D. grandis, D. indicus, D. frigens, D. saxicola, D. marmoris, D. deserti, D. geothermalis* and *D. murrayi*; the latter two are also thermophilic.

History

D. *radiodurans* was discovered in 1956 by Arthur W. Anderson at the Oregon Agricultural Experiment Station in Corvallis, Oregon. Experiments were being performed to determine if canned food could be sterilized using high doses of gamma radiation. A tin of meat was exposed to a dose of radiation that was thought to kill all known forms of life, but the meat subsequently spoiled, and D. *radiodurans* was isolated. The complete DNA sequence of D. *radiodurans* was published in 1999 by TIGR. A detailed annotation and analysis of the genome appeared in 2001. The sequenced strain was ATCC BAA-816. *Deinococcus radiodurans* has a unique quality in which it can repair DNA both single and double stranded. When a mutation is apparent to the cell it brings it into a compartmental ring like structure where the DNA is repaired and then is able to fuse the nucleoids from the outside of the compartment with the damaged DNA.

Description

D. *radiodurans* is a rather large spherical bacterium, with a diameter of 1.5 to 3.5 μm. Four cells normally stick together, forming a tetrad. The bacteria are easily cultured and do not appear to cause disease. Colonies are smooth, convex, and pink to red in colour. The cells stain gram positive, although its cell envelope is unusual and is reminiscent of the cell walls of gram negative bacteria. D. *radiodurans* does not form endospores and is nonmotile. It is an obligate aerobic chemoorganoheterotroph, i.e. it uses oxygen to derive energy from organic compounds in its environment. It is often found in habitats rich in organic materials, such as soil, feces, meat, or sewage, but has also been isolated from dried foods, room dust, medical instruments and textiles.

It is extremely resistant to ionizing radiation, ultraviolet light, desiccation, oxidizing and electrophilic agents. Its genome consists of two circular chromosomes, one 2.65 million base pairs long and the other 412,000 base pairs long, as well as a megaplasmid of 177,000 base pairs and a plasmid of 46,000 base pairs. It has about 3,195 genes. In its stationary phase each bacterial cell contains four copies of this genome; when rapidly multiplying, each bacterium contains 8-10 copies of the genome.

Ionizing Radiation Resistance

D. *radiodurans* is capable of withstanding an acute dose of 5,000 Gy of ionizing radiation with almost no loss of viability, and an acute dose of 15,000 Gy with 37% viability. A dose of 5,000 Gy is estimated to introduce several hundred double strand breaks (DSBs) into the organism's DNA (~0.005 DSB/Gy/Mbp (haploid genome)). For comparison, a chest X-ray or Apollo mission involves about 1 milligray, 5 Gy can kill a human, 200-800 Gy will kill *E. coli,* and over 4,000 Gy will kill the radiation-resistant tardigrade.

Several bacteria of comparable radioresistance are now known, including some species of the genus *Chroococcidiopsis* (phylum cyanobacteria) and some species of *Rubrobacter* (phylum actinobacteria); among the archaea, the species *Thermococcus gammatolerans* shows comparable radioresistance. *Deinocuccus radiodurans* also has a unique ability to repair damaged DNA. It isolates the damaged segments in a controlled area and repairs it. This bacteria can also repair many small fragments from an entire chromosome.

Ionizing Radiation Resistance Mechanisms

Deinococcus accomplishes its resistance to radiation by having multiple copies of its genome and rapid DNA repair mechanisms. It usually repairs breaks in its chromosomes within 12–24 hours through a 2-step process.

First, *D. radiodurans* reconnects some chromosome fragments through a process called single-strand annealing. In the second step, a protein mends double-strand breaks through homologous recombination. This process does not introduce any more mutations than a normal round of replication would.

A persistent question regarding *D. radiodurans* is how such a high degree of radioresistance could evolve.

Natural background radiation levels are very low—in most places, on the order of 0.4 mGy per year, and the highest known background radiation, near Ramsar, Iran is only 260 mGy per year. With naturally-occurring background radiation levels so low, organisms evolving mechanisms specifically to ward off the effects of high radiation are unlikely.

Valerie Mattimore and John R. Battista of Louisiana State University have suggested that the radioresistance of *D. radiodurans* is simply a side-effect of a mechanism for dealing with prolonged cellular desiccation (dryness). To support this hypothesis, they performed an experiment in which they demonstrated that mutant strains of *D. radiodurans* which are highly susceptible to damage from ionizing radiation are also highly susceptible to damage from prolonged desiccation, while the wild type strain is resistant to both. In addition to DNA repair, *D. radiodurans* use LEA proteins (Late Embryogenesis Abundant proteins) expression to protect against desiccation.

Scanning electron microscopy analysis has shown that DNA in *D. radiodurans* is organized into tightly packed toroids, which may facilitate DNA repair.

A team of Croatian and French researchers led by Miroslav Radman have bombarded *D. radiodurans* to study the mechanism of DNA repair. At least two copies of the genome, with random DNA breaks, can form DNA fragments through annealing. Partially overlapping fragments are then used for synthesis of homologous regions through a moving D-loop that can continue extension until they find complementary partner strands. In the final step there is crossover by means of RecA-dependent homologous recombination.

Michael Daly has suggested that the bacterium uses manganese complexes as antioxidants to protect itself against radiation damage. In 2007 his team showed that high intracellular levels of manganese(II) in *D. radiodurans* protect proteins from being oxidized by radiation, and proposed the idea that "protein, rather than DNA, is the principal target of the biological action of [ionizing radiation] in sensitive bacteria, and extreme resistance in Mn-accumulating bacteria is based on protein protection".

A team of Russian and American scientists proposed that the radioresistance of *D. radiodurans* had a Martian origin. Evolution of the microorganism could have taken place on the Martian surface until it was delivered to Earth on a meteorite. However, apart from its resistance to radiation, Deinococcus is genetically and biochemically very similar to other terrestrial life forms, arguing against an extraterrestrial origin.

In 2009 it was reported that nitric oxide plays an important role in the bacteria's recovery from radiation exposure: the gas is required for division and proliferation after DNA damage has been repaired. A gene was described that increases nitric oxide production after UV radiation, and in the absence of this gene the bacteria were still able to repair DNA damage but would not grow.

Applications

Deinococcus has been genetically engineered for use in bioremediation to consume and digest solvents and heavy metals, even in a highly radioactive site. For example, the bacterial mercuric reductase gene has been cloned from *Escherichia coli* into *Deinococcus* to detoxify the ionic mercury residue frequently found in radioactive waste generated from nuclear weapons manufacture. Those researchers developed a strain of *Deinococcus* that could detoxify both mercury and toluene in mixed radioactive wastes. The Craig Venter Institute has used a system derived from the rapid DNA repair mechanisms of *D. radiodurans* to assemble synthetic DNA fragments into chromosomes, with the ultimate goal of producing a synthetic organism they call *Mycoplasma laboratorium*.

In 2003, U.S. scientists demonstrated that *D. radiodurans* could be used as a means of information storage that might survive a nuclear catastrophe. They translated the song *It's a Small World* into a series of DNA segments 150 base pairs long, inserted these into the bacteria, and were able to retrieve them without errors 100 bacterial generations later.Thermococcus gammatolerans.

Thermococcus gammatolerans is an archaea extremophile and the most radiation resistant known organism. Recently discovered in a submarine hydrothermal vent in the Guaymas Basin about 2,000 meters deep off the coast of California.

Thermococcus gammatolerans sp. nov thrives in temperatures between 55-95 °C with an optimum development at approximately 88 °C. The optimal growth pH is 6, favoring the presence of sulfur (S), which is reduced to hydrogen sulfide (H2S). It is the organism with the strongest resistance to radiation, supporting a radiation of gamma rays from 30 KGy.

It belongs to the family of Thermococci they are a group of microorganisms in the Archaea domain and phylum Euryarchaeota. The Thermococci live in extremely hot environments such as hydrothermal vents with a growth optimum temperature above 80 °C. Thermococcus and Pyrococcus (literally "ball of fire") are both chemoorganotrophic anaerobic required.

Thermococcus prefer 70-95 °C, whereas the range is better Pyrococcus 70-100 °C. The resistance to ionizing radiation of T. Gammatolerans is enormous, while a dose of 10 Gy is sufficient to kill a human being, and a dose of 60 Gy is able to kill all cells in a colony of E. coli, the Thermococcus gammatolerans sp. nov can withstand an instantaneous dose of up to 5,000 Gy with no loss of viability and doses of up to 30 000 Gy.

History

The gammatolerans Thermococcus sp. nov, was discovered in 2003 on samples collected from a hydrothermal chimney collected at the Guaymas Basin about 2,000 meters deep off the coast of California. (27 degrees 01 'N, 111° 24' W).

Mechanisms of Resistance to Radiation

Unlike other organisms Thermococcus gammatolerans cell survival is not altered by the changing conditions in its growth phase, but the lack of ideal conditions and nutrients decreases its radioresistance. The system of chromosomal DNA repair shows that cells in stationary phase of growth reconstitute the DNA more rapidly than cells in exponential growth phase. T. Gammatolerans can slowly or quickly rebuild damaged chromosomes without loss of viability.

Applications

It has been studied its application to the development of new enzymatic markers are resistant to high temperatures. Their application in the study of carcinogenesis and the study of the development of mitochondrial diseases. Has been speculated that DNA repair mechanisms of T. gammatolerans could be incorporated into the genome of higher species in order to improve DNA repair and reduce cellular aging.

Radiotrophic Fungus

Radiotrophic fungi are fungi which appear to use the pigment melanin to convert gamma radiation into chemical energy for growth. This proposed mechanism may be similar to anabolic pathways for the synthesis of reduced organic carbon (e.g., carbohydrates) in phototrophic organisms, which capture photons from visible light with pigments such as chlorophyll whose energy is then used in photolysis of water to generate usable chemical energy (as ATP) in photophosphorylation of photosynthesis.

However, whether melanin-containing fungi employ a similar multi-step pathway as photosynthesis, or some chemosynthesis pathways, is unknown. These were first discovered in 2007 as black molds growing inside and around the Chernobyl Nuclear Power Plant. Research at the Albert Einstein College of Medicine showed that three melanin-containing fungi, *Cladosporium sphaerospermum*, *Wangiella dermatitidis*, and *Cryptococcus neoformans*, increased in biomass and accumulated acetate faster in an environment in which the radiation level was 500 times higher than in the normal environment. Exposure of *C. neoformans* cells to these radiation levels rapidly (within 20–40 minutes of exposure) altered the chemical properties of its melanin and increased melanin-mediated rates of electron transfer (measured as reduction of ferricyanide by NADH) 3 to 4-fold compared with unexposed cells. Similar effects on melanin electron-transport capability were observed by the authors after exposure to non-ionizing radiation, suggesting that melanotic fungi might also be able to use light or heat radiation for growth. However, melanization may come at some metabolic cost to the fungal cells: in the absence of radiation, some non-melanized fungi (that had been mutated in the melanin pathway) grew faster than their melanized counterparts. Limited uptake of nutrients due to the melanin molecules in the fungal cell wall or toxic intermediates formed in melanin biosynthesis have been suggested to contribute to this phenomenon.

Relation of Bacteria to Agriculture

The Silo

In the management of a silo the farmer has undoubtedly another great bacteriological problem. In the attempt to preserve his summer-grown food for the winter use of his animals, he is hindered by the activity of common bacteria. If the food is kept moist, it is sure to undergo decomposition and be ruined in a short time as animal food. The farmer finds it necessary, therefore, to dry some kinds of foods, like hay. While he can thus preserve some foods, others can not be so treated.

Much of the rank growth of the farm, like cornstalks, is good food while it is fresh, but is of little value when dried. The farmer has from experience and observation discovered a method of managing bacterial growth which enables him to avoid their ordinary evil effects. This is by the use of the silo. The silo is a large, heavily built box, which is open only at the top. In the silo the green food is packed tightly, and when full all access of air is excluded, except at its surface. Under these conditions the food remains moist, but nevertheless does not undergo its ordinary fermentations and ;putrefactions, and may be preserved for months without being ruined. The food in such a silo may be taken out months after it is packed, and will still be found to be in good condition for food. It is true that it has changed its character somewhat, but it is not decayed, and is eagerly eaten by cattle.

We are yet very ignorant of the nature of the changes which occur in the food while in the silo. The food is not preserved from fermentation. When the silo is packed slowly, a very decided fermentation occurs by which the mass is raised to a high temperature (1400 F. to 160° F.).

This heating is produced by certain species of bacteria which grow readily even at this high temperature. The fermentation uses up the air in the silo to a certain extent and produces a settling of the material which still further excludes air. The first fermentation soon ceases, and afterward only slow changes occur. Certain acid-producing bacteria after a little begin to grow slowly, and in time the silage is rendered somewhat sour by the production of acetic acid.

But the exclusion of air, the close packing, and the small amount of moisture appear to prevent the growth of the common putrefactive bacteria, and the silage remains good for a long time. In other methods of filling the silo, the food is very quickly packed and densely crowded together so as to exclude as much air as possible from the beginning. Under these conditions the lack of moisture and air prevents fermentative action very largely. Only certain acid-producing organisms grow, and these very slowly.

The essential result in either case is that the common putrefactive bacteria are prevented from growing, probably by lack of sufficient oxygen and moisture, and thus the decay is prevented. The closely packed food offers just the same unfavourable condition for the growth of common putrefactive bacteria that we have already seen offered by the hard-pressed cheese, and the bacteria growth is in the same way held in check. Our knowledge of the matter is as yet very slight, but we do know enough to understand that the successful management of a silo is dependent upon the manipulation of bacteria.

The Fertility of The Soil

The farmer's sole duty is to extract food from the soil. This he does either directly by raising crops, or indirectly by raising animals which feed upon the products of the soil. In either case the fertility of the soil is the fundamental factor in his success. This fertility is a gift to him from the bacteria.

Even in the first formation of soil he is in a measure dependent upon bacteria. Soil, as is well known, is produced in large part by the crumbling of the rocks into powder. This crumbling we generally call weathering, and regard it as due to the effect of moisture and cold upon the rocks, together with the oxidizing action of the air. Doubtless this is true, and the weathering action is largely a physical and chemical one. Nevertheless, in this fundamental process of rock disintegration bacterial action plays a part, though perhaps a small one. Some species of bacteria, as we have seen, can live upon very simple foods, finding in free nitrogen and carbonates sufficiently highly complex material for their life. These organisms appear to grow on the bare surface of rocks, assimilating nitrogen from the air, and carbon from some widely diffused carbonates or from the CO, in the air. Their secreted products of an acid nature help to soften the rocks, and thus aid in performing the first step in weathering.

The soil is not, however, all made up of disintegrated rocks. It contains, besides, various ingredients which combine to make it fertile. Among these are various sulphates which form important parts of plant foods. These sulphates appear to be formed, in part, at least, by bacterial agency. The decomposition of proteids gives rise, among other things, to hydrogen sulphide (H,S). This gas, which is of common occurrence in the atmosphere, is oxidized by bacterial growth into sulphuric acid, and this is the basis of part of the soil sulphates. The deposition of iron phosphates and iron silicates is probably also in a measure aided by bacterial action. All of these processes are factors in the formation of soil. Beyond much question the rock disintegration which occurs everywhere in Nature is chiefly the result of physical and chemical changes, but there is reason for believing that the physical and chemical processes are, to a slight extent at least, assisted by bacterial life.

A more important factor of soil fertility is its nitrogen content, without which it is completely barren. The origin of these nitrogen ingredients has been more or less of a puzzle. Fertile soil everywhere contains nitrates and other nitrogen compounds, and in certain parts of the world there are large accumulations of these compounds, like the nitrate beds of Chili. That they have come ultimately from the free atmospheric nitrogen seems certain, and various attempts have been made to explain a method of this nitrogen fixation. It has been suggested that electrical discharges in the air may form nitric acid, which would readily then unite with soil ingredients to form nitrates. There is little reason, however, for believing this to be a very important factor. But in the soil bacteria we find undoubtedly an efficient agency in this nitrogen fixation. As already seen, the bacteria are able to seize the free atmospheric nitrogen, converting it into nitrites and nitrates. We have also learned that they can act in connection with legumes and some other plants, enabling them to fix atmospheric nitrogen and store it in their roots. By these two means the nitrogen ingredient in the soil is prevented from becoming exhausted by the processes of dissipation constantly going on. Further, by some such agency must we imagine the original nitrogen soil ingredient to have been derived. Such an organic agency is the only one yet discerned which appears to have been efficient in furnishing virgin soil with its nitrates, and we must therefore look upon bacteria as essential to the original fertility of the soil.

But in another direction still does the farmer depend directly upon bacteria. The most important factor in the fertility of the soil is the part of it called humus. This humus is very complex, and never alike in different soils. It contains nitrogen compounds in abundance, together with sulphates, phosphates, sugar, and many other substances. It is this which makes the garden soil different from sand, or the rich soil different from the sterile soil. If the soil is cultivated year after year, its food ingredients are slowly but surely exhausted. Something is taken from the humus each year, and unless this be replaced the soil ceases to be able to support life. To keep up a constant yield from the soil the farmer understands that he must apply fertilizers more or less constantly.

This application of fertilizers is simply feeding the crops. Some of these fertilizers the farmer purchases, and knows little or nothing as to their origin. The most common method of feeding the crops is, however, by the use of ordinary barnyard manure.

The reason why this material contains plant food we can understand, since it is made of the undigested part of food, together with all the urea and other excretions of animals, and contains, therefore, besides various minerals, all of the nitrogenous waste of animal life. These secretions are not at first fit for plant food. The farmer has learned by experience that such excretions, before they are of any use on his fields, must undergo a process of slow change, which is sometimes called ripening.

Fresh manure is sometimes used on the fields, but it is only made use of by the plants after the ripening process has occurred. Fresh animal excretions are of little or no value as a fertilizer. The farmer, therefore, commonly allows it to remain in heaps for some time, and it undergoes a slow change, which gradually converts it into a condition in which it can be used by plants. This ripening is readily explained by the facts already considered. The fresh animal secretions consist of various highly complex compounds of nitrogen, and the ripening is a process of their decomposition.

The proteids are broken to pieces, and their nitrogen elements reduced to the form of nitrates, leucin, etc., or even to ammonia or free nitrogen. Further, a second process occurs, the process of oxidation of these nitrogen compounds already noticed, and the ammonia and nitrites resulting from the decomposition are built into nitrates. In short, in this ripening manure the processes noticed in the first part of this chapter are taking place, by which the complex nitrogenous bodies are first reduced and then oxidized to form plant food. The ripening of manure is both an analytical and a synthetical process. By the analysis, proteids and other bodies are broken into very simple compounds, some of them, indeed, being dissipated into the air, but other portions are retained and then oxidized, and these latter become the real fertilizing materials. Through the agency of bacteria the compost heap thus becomes the great source of plant food to the farmer. Into this compost heap he throws garbage, straw, vegetable and animal substances in general, or any organic refuse which may be at hand. The various bacteria seize it all, and cause the decomposition which converts it into plant food again. The rotting of the compost heap is thus a gigantic cultivation of bacteria.

This knowledge of the ripening process is further teaching the farmer how to prevent waste. In the ordinary decomposition of the compost heap not an inconsiderable portion of the nitrogen is lost in the air by dissipation as ammonia or free nitrogen. Even his nitrates may be thus lost by bacterial action. This portion is lost to the farmer completely, and he can only hope to replace it either by purchasing nitrates in the form of commercial fertilizers, or by reclaiming it from the air by the use of the bacterial agencies already noticed. With the knowledge now at his command he is learning to prevent this waste. In the decomposition one large factor of loss is the ammonia, which, being a gas, is readily dissipated into the air.

Knowing this common result of bacterial action, the scientist has told the farmer that, by adding certain common chemicals to his decomposing manure heap, chemicals which will readily unite with ammonia, he may retain most of the nitrogen in this heap in the form of ammonia salts, which, once formed, no longer show a tendency to dissipate into the air. Ordinary gypsum, or superphosphates, or plaster will readily unite with ammonia, and these added to the manure heap largely counteract the tendency of the nitrogen to waste, thus enabling the farmer to put back into his soil most of the nitrogen which was extracted from it by his crops and then used by his stock. His vegetable crops raise the nitrates into proteids. His animals feed upon the proteids, and perform his work or furnish him with milk. Then his bacteria stock take the excreted or refuse nitrogen, and in his manure heap turn it back again into nitrates ready to begin the circle once more. This might go on almost indefinitely were it not for two facts : the farmer sends nitrogenous material off his farm in the milk or grains or other nitrogenous products which he sells, and the decomposition processes, as we have seen, dissipate some of the nitrogen into the air as free nitrogen.

To meet this emergency and loss the farmer has another method of enriching the soil, again depending upon bacteria. This is the so-called green manuring. Here certain plants which seize nitrogen from the air are cultivated upon the field to be fertilized, and, instead of harvesting a crop, it is ploughed into the soil. Or perhaps the tops may be harvested, the rest being ploughed into the soil. The vegetable material thus ploughed in lies over a season and enriches the soil. Here the bacteria of the soil come into play in several directions. First, if the crop sowed be a legume, the soil bacteria assist it to seize the nitrogen from the air. The only plants which are of use in this green manuring are those which can, through the agency of bacteria, obtain nitrogen from the air and store it in their roots. Second, after the crop is ploughed into the soil various decomposing bacteria seize upon it, pulling the compounds to pieces.

The carbon is largely dissipated into the air as carbonic dioxide, where the next generation of plants can get hold of it. The minerals and the nitrogen remain in the soil. The nitrogenous portions go through the same series of decomposition and synthetical changes already described, and thus eventually the nitrogen seized from the air by the combined action of the legumes and the bacteria is converted into nitrates, and will serve for food for the next set of plants grown on the same soil. Here is thus a practical method of 'using the nitrogen assimilation powers of bacteria, and reclaiming nitrogen from the air to replace that which has been lost.

Thus it is that the farmer's nitrogen problem of the fertile soil appears to resolve itself into a proper handling of bacteria. These organisms have stocked his soil in the first place. They convert all of his compost heap wastes into simple bodies, some of which are changed into plant foods, while others are at the same time lost. Lastly, they may be made to reclaim this lost nitrogen, and the farmer, so soon as he has requisite knowledge of these facts, will be able to keep within his control the supply of this important element. The continued fertility of the soil is thus a gift from the bacteria.

Sustainable Agriculture Focus on Issues Facing Farmers and Producers

The issues that are important to farmers across the country include stewardship, profitability, sustainability and the health of their community. And those are the topics to be covered in Kansas in an upcoming agriculture roundup.

The practice of sustainable agriculture is built upon soil fertility and the protection of soil health. For centuries, farmers around the world have been employing these basic techniques to keep their soil productive. Early in colonial American history, George Washington was using cover crops and manure applications on his farm. American farmers, who did not carefully tend their soil, eventually "wore it out." The Westward Expansion partly reflected the lowered productivity of eastern lands and the search for new farmland farther west.

The first publications from the Kansas State Agricultural College appeared in the late 1880's, at a time when Kansas soils had been tilled for less than fifty years. Soil scientists were well aware at that time that the long term sustainability of farming depended upon the use of legumes for fertility and the addition of organic matter for tilth. As Kansas farmers began to report declining fertility levels in the early 1900's, scientists cautioned that farmers must not rely solely upon the natural fertility of the prairie soils.

Between 1900 and 1950, Kansas State researchers guided farmers in the use of legumes and the preservation of soil organic matter. With the advent of inexpensive nitrogen fertilizers after World War II, traditional soil management techniques took a back seat to the use of commercial products that were easy to use and often provided greater economic returns in the short-term.

At the beginning of the twenty-first century, farmers are becoming increasingly conscious of the importance of soil health, water quality, and energy conservation. The rising cost of nitrogen fertilizers has revived interest in nitrogen-fixing legumes. Excess phosphorus levels in surface water indicate a need to emphasize soil conservation and the careful use of manure resources. Traditional soil management practices continue to be vital for the sustainable agriculturist and are regaining an audience with conventional agriculture.

This review of Kansas State University soil publications profiles the research and recommendations of Kansas scientists during the early to mid-twentieth century. Generally, recommended practices such as crop rotations, manure use, cover crops, and other sources of fertility are considered including the shift to commercial fertilizers in the 1950's. The special consideration of western Kansas soils is treated separately. Limited rainfall in western Kansas affects the research conducted in that part of the state and alters the use of traditional soil health practices.

Soil Health

Within every aspect of Kansas agriculture, healthy soil is a key element. Its structure and fertility provide the basis for all crop and livestock production. Throughout the historical publications of Kansas State University from the late 1800's to the mid-1900's, researchers and educators have been concerned with the protection of this vital resource. In their 1918 publication, Soil Fertility, L.E. Call and R. I Throckmorton caution the Kansas farmer. "The soil is the most important source of wealth in an agricultural state. If it is maintained in a high state of productivity, by wise systems of soil management, the people prosper. If its fertility is wasted through careless methods of farming, both the farmer and the state suffer".

Forty years later, in 1956, Orville Bidwell echoes this same message with its promise of a precarious wealth. "Unlike most other resources, soil is inexhaustible if properly managed." Bidwell recounts the variety of Kansas soils each with a different waterholding capacity, permeability, response to fertilizers, and susceptibility to erosion.

Decline of Kansas Soils

Throughout many of the earliest publications, the authors are clearly concerned about the declining condition of Kansas soils. By the early 1900's, much of the rich prairie soil in eastern Kansas had been farmed for 50 years. A 1903 publication from the veterinary department of the Kansas State Experiment Station states, "The fertility of the soil of the Middle states and the West is being rapidly diminished and if means are not taken to prevent it, the time is not far distant when it will be as necessary to apply artificial fertilizers to the soil as it is now in the East".

A few years later, chemists at the agricultural experiment station raised the same concern. "In the early history of Kansas no attention was paid to the composition of its soils except to boast of their inexhaustible fertility. The voice of the chemist has been lifted constantly, warning the people that this idea of possession of a fertility that is practically limitless is a delusion that can lead only to squandering of our natural resources, and to leaving posterity handicapped in the struggle for existence. Today he is seeing his warnings justified. People in many localities of the eastern part of the state are making inquiry concerning chemical analysis of their soils with reference to learning what fertilizers should be applied and to what crops their soils are best adapted".

By 1918, L.E. Call and R. I. Throckmorton attempted to put dollar figures to the losses in soil fertility. They estimated that the plant food removed from Kansas soils by wheat crops over the previous fifty-five years equalled a value of more than seven hundred million dollars. Even the wheat straw, which was regularly burned or wasted, had a value in plant nutrients of more than twelve million dollars. The majority of the wheat products were both milled and eaten outside the state which Call and Throckmorton equated with the export of soil fertility.

Call and Throckmorton credited the declining productivity of Kansas soils to five factors: depletion of soil organic matter, failure to grow enough acres of leguminous crops for nitrogen fixation, depletion of mineral nutrients, the lack of proper crop rotations, and the erosion of fertile topsoil. These five factors are the subject of nearly every soils publication prior to the advent of inexpensive nitrogen fertilizers in the mid-1900's. They continue to be the basis of soil health for every farming system regardless of the use of commercial fertilizers.

Soil Organic Matter

The early Kansas State publications recognized the importance of soil organic matter to soil health. Replenishment of soil organic matter is the most basic step in addressing a number of other issues. According to Call and Throckmorton, organic matter holds the soil's store of nitrogen, provides good tilth, holds moisture, and provides food for the bacteria that make nutrients available to plants.

Soil studies in western Kansas spanning over 30 years of crop production give mention to the importance of organic matter. Progress reports in 1943 and 1957 indicate that as the organic carbon content of the soil has decreased, more power is needed for tillage, water intake is decreased, seedling emergence and root growth are hindered. The 1943 report also mentions that the presence of coarse, fibrous organic matter helped reduce wind erosion. Numerous publications cite frequent plowing and intensive cultivation as culprits in the loss of organic matter. Row crops depleted the soil more quickly than small grain crops. Without a regular practice of restoring organic matter, the levels of carbon in the soil would drop, threatening fertility and soil structure.

The 1918 publication, Soil Fertility, promotes barnyard manure applied to the soil as a primary source of organic matter. However, when sufficient manure is not available, the farmer should grow a crop to plow under to supply organic matter. These green manure crops may be legumes such as alfalfa, cowpeas, soybeans, clover, or sweet clover. Legumes would capture atmospheric nitrogen and fix it in the soil in addition to supplying organic matter. Non-legume crops such as rye, buckwheat, sorghum, and turnips were also cited as possible sources for organic matter.

Although most of the methods for increasing organic matter focus on the addition of green manure crops or barnyard manure, the conservation of all organic matter is addressed. Throckmorton and Call state that most wheat straw is either burned or destroyed after threshing. They recommend its use as feed and bedding for livestock. Eventually the soiled bedding and any manure would be returned to the soil. They also suggest use of the straw as a surface mulch on wheat during the winter at a rate of 1 - 1.5 tons per acre.

In 1962, farmers were again cautioned not to waste the organic matter provided by straw or stubble. In an effort to control weeds and increase yields, burning of wheat stubble had become a common practice. A series of studies in western Kansas produced data showing that stubble burning did not increase yields for subsequent crops. Not only did it present increased potential for wind erosion, it also decreased the soil's ability to absorb water.

Nitrogen and Organic Matter

The supply of nitrogen in the soil was closely linked to the organic matter content. The practice of returning organic matter to the soil in the form of leguminous plants or animal manures was also the method for supplying nitrogen prior to the use of commercial nitrogen fertilizer. Legume crops, grown in rotation with other crops, could be worked back into the soil as a green manure crops or the legume hay crop was fed to livestock and their manure was returned to the soil.

Off-farm sources of nitrogen in the early 1900's included waste materials from the packing houses and inorganic compounds such as saltpeter, which was mined in Chili, and manufactured compounds of ammonium sulphate or calcium cyanamide. Researchers noted that "the purchase of nitrogenous fertilizers should be limited to the meeting of special requirements of certain conditions or crops...Nitrogen...is the (nutrient) most cheaply restored, since by the cooperation of clovers, alfalfa, peas, beans, and other legumes with bacteria that grow upon their roots the abundant nitrogen of the air in the pores of the soil is brought into organic combination. This means of adding nitrogen to a soil must never be lost to view...".

Through the first half of the twentieth century, researchers and farmers sought to understand the best methods for capturing nitrogen with legumes especially when those crops were being used for other purposes on the farm. In 1918, researchers were concerned that the nitrogen found within the crop roots and stubble might not be significant compared with the amount removed in a hay crop. They hypothesized that a certain amount of leaf loss during the haying process might be returning some nitrogen to the soil. Still they cautioned that the best practice was to feed the hay on the farm and return all manure to the soil.

The 1939 publication on fertility studies at the Manhattan experiment station beginning in 1910 gave evidence that soil nitrogen could be increased even when the hay was removed. The average increase in the 5-9 year old hay plots was twice that for the 1-4 year plots. The researchers also found that the residual effects of the "nitrogen accumulating capacity" remained for eight to nine years following the breaking of the alfalfa sod that had been in alfalfa for more than two years.

But nitrogen was not the only concern when hay was taken off the farm. "If alfalfa is sold off the land it is one of the most soil-exhausting crops raised, while if it is fed on the farm, and the manure produced applied to the land, it is a conserver of fertility. The same argument applies to clover". Potassium, phosphorus, calcium and other mineral elements are taken up by the plants and must be cycled back to the soil as green manure or barnyard manure. When hay is sold off the farm, the phosphorus and potassium are more rapidly depleted than with a continuous grain crop.

Regardless of the use of green manure crops, any farm with livestock had a source of organic matter and nutrients in animal manure if it was used wisely. "If the livestock farmer properly saves and utilizes his manure he can maintain his soil in a high state of productivity, but the livestockman who feeds his cattle in woodlots along the banks of streams, and so wastes his manure, usually depletes the fertility of his soil more rapidly than the man producing grain only. It was important to manage manure so that nutrients were not lost before they were cycled back to the soil. The seepage of liquid waste or urine, leaching of manure by rain and runoff water, and the "decay" of manure solids and the subsequent loss of nitrogen are all a result of poor handling techniques. Throckmorton and Call recommend that the most practical method of manure handling would be to feed the stock on the cultivated fields so that the manure is scattered by the animals and the nutrients are retained by the soil. In any case, the manure should be returned to the soil as soon as possible and long term, open storage of six months or more should be avoided.

Manure could also be used sparingly as a top dressing on corn, kafir, or winter wheat where it would act as mulch to retain moisture. If there was not a large supply of manure, it was deemed better to apply the available supply lightly to more acres rather than a heavy application on just a few acres.

As a percentage of volume, the nutrients present in manure are small. "The figures for the fertilizing constituents are always low, but they are present in readily available form and the accompanying organic matter has itself a highly beneficial effect on the land. Even with these low percentages the total amount of plant food in the manure produced on a farm reaches very significant quantities".

Phosphorus

Fertilizers were commonly used to provide phosphorus and other minerals long before nitrogen fertilizers were in general use. By 1918, the soils in eastern and southeast Kansas had such low stores of phosphorus that it was profitable to purchase phosphorus fertilizers. Grain farms lost phosphorus the most rapidly when it was exported off the farm with the grain. In order to retain the mineral, the farmer had to feed the grain to livestock and apply the manure to the croplands or else import a supply of phosphorus.

Alfalfa and other deep-rooted crops could be used to collect phosphorus from the subsoil. The nutrients then needed to be cycled back into the upper levels of the soil as either a green manure crop or manure from animals fed on the hay..

Commercial sources of phosphorus included bone, basic slag, rock phosphate, and apatite. Bones were a valuable product rich in phosphorus and nitrogen. They could be ground raw but they were more commonly steamed before grinding. This processing made grinding easier, concentrated the phosphorus slightly and increased the availability to plants.

Basic slag was a by-product from a particular method of iron refinement. Minerals removed from certain iron ore contained high amounts of phosphate, which could be ground for use as a fertilizer. This was not a common product in America, being chiefly available in Europe.

Primary sources of rock phosphate were found in Florida, South Carolina, and Tennessee during the early part of the twentieth century. Although it was most often used to produce superphosphate, it could be finely ground and used raw. Superphosphate was produced by treating the raw phosphate with sulphuric acid and was in use to some extent throughout the twentieth century.

Although superphosphate was more readily available to the plant, Kansas State researchers advised that the raw phosphate was usually a wiser choice. "In the application of phosphate fertilizers the farmer naturally expects and desires immediate results, which are secured by the use of superphosphate, but at the same time if larger quantities of phosphates can be applied in other less soluble and available forms at the same expenditure, the ultimate value of the investment may be much greater, as the phosphorus will remain in the soil and be rendered available by slow natural processes.

There was also indicated a positive relationship between the bacteria found in organic matter and the availability of phosphorus from raw phosphate. "The cheapest source of phosphorus is ground rock phosphate, and in this form it will be available for the use of crops if the soil is well supplied with organic matter from farm manures and legumes.

Apatite, a crystal found in granite, was abundantly available in Canada. It needed to be converted to superphosphate to be used as a fertilizer.

Potassium

Potassium occurs in the mineral or rock portion of the soil. Plant roots are able to access potassium directly from tiny soil particles, chiefly silt and clay. Potassium is also present in decaying crop residues and organic matter, indicating the need, once again, for returning plant materials to the soil. A traditional source of additional potassium was hardwood ashes. Muriate of potash, a processed form of potassium in wide use today, was also available early in the twentieth century but due to the chlorine content of the compound, its use was restricted with certain crops.

Calcium (Lime)

Calcium, while an essential element for plant growth, was usually not deficient in Kansas soils. However, its application in the form of lime, was a common soil amendment. "(Calcium) is generally present in all cultivated soils in sufficient quantity to supply fully the need of the plant. Yet even where this is the case, the soil may be greatly in need of liming. Lime is used, therefore, as a soil amendment not so much for its effect directly on the plant as for its effect on the soil, which indirectly affects the plant. Soils that are low in lime are said to be sour". The acidity of these soils could be neutralized by the application of calcium compounds.

A 1918 publication recommended two tons of ground limestone per acre followed by one or two tons every five to six years thereafter. The limestone should be worked into the bare ground six months to a year prior to seeding alfalfa or clovers. The effects of the limestone are slow and gradual but long lasting. The bacteria that live on the roots of alfalfa, sweet clover, and clover thrive in more alkaline soils so these crops respond well to liming. Crops considered less sensitive to acid soils were corn, wheat, timothy, and oats.

The change in soil pH improved the availability of phosphorus and potassium as well as improving the texture of the soil. "It is a well-known fact that soils well stocked with a supply of lime will be more productive under the same conditions of plant-food content than soils not so stocked...While soils use very small amounts of lime as compared with phosphorus and potassium, yet the presence of a relatively large supply of lime insures crop production. In the words of Hilgard, 'A lime country is a rich country'".

Soil Bacteria

Although the exact relationship did not seem to be clear, researchers in the early part of the twentieth century were writing about a correlation between humus or organic matter, bacteria, and fertility. A series of studies beginning in the 1890's showed crop yields in "direct proportion" to the bacterial content of the fields. In this same publication, the authors speculate that fertility depends to a large extent on bacterial activity and that by manipulating the bacteria of the soil, one might avoid the need for artificial fertilization. They proposed additional experiments to find ways to increase soil bacteria.

A few years later, Walter King and Charles Doryland report on their studies of soil bacteria with this same basic assumption. Assuming that bacterial activity indicated increased soil fertility, they set about collecting samples from various soil depths on plots representing a variety of tillage practices. They admitted tremendous variability in their data, which was collected over a period of only three months from March to June 1908. Nevertheless, they concluded that deep plowing, conveniently the tillage practice most commonly used at that time, increased bacteria levels and bacterial activity and decreased denitrification. They noted that bacterial activity increased with the temperature of the soil and decreased when the soil became saturated with moisture. Different species were predominate at different times and activity seemed to rise and fall with a regularity independent of moisture and temperature.

Analysis of the Soil

As farmers began to experience decreasing yields on exhausted soil, they began to show interest in chemical analysis of their soil. Inexpensive, reliable soil tests were not yet available in the early part of the century. Throckmorton and Call noted that although they could determine the nutrient needs of a crop and the chemical analysis of the soil, their current methods did not tell them what nutrients were available to the crop. They felt that available nutrients "fluctuate greatly" depending upon the total nutrients in the soil, the organic matter content, the weather, cultivation methods and the current crop. Consequently, a chemical analysis would only give a farmer a general outline. Probably the greatest deterrent to using a chemical analysis was the expense which made it impractical for individual farmers.

In 1909, the Extension Station council authorized the Chemistry Department to collect and analyze typical soils from across Kansas. This body of work did give farmers an idea of the basic properties of their soil and requirements of common crops even if it could not provide specific answers regarding fertility.

In 1910, the Chemistry Department cautioned that chemical analysis of the soil was not a reliable indicator regarding crop needs. They recommended that the farmer should test his soil by raising small crop plots that had been "fractionally fertilized" in order to determine the optimal rates. The complexities of variable rate plots involving a significant amount of land and more than one growing season probably doomed this testing method for on-farm use.

More practical advice involved a farmer's observational skills. "Chemical and physical investigation of soils...must be supplemented or...replaced by observations upon the natural growth of trees, shrubs, grasses or weeds upon the soil, and by experiments in the production of plants or crops upon it. Let organic nature answer the question, What is this soil good for? Observations concerning the natural plant growth upon a soil have always been used by practical men in judging of its value...This means of gaining an insight into soil values is one that, while used from time immemorial, is worthy of more extended study and application

Commercial Fertilizers

Throughout the Kansas State publications from the first half of the twentieth century, farmers are cautioned about reliance on chemical fertilizers. "It should not be forgotten...that barnyard manure, because of its content of organic matter in a state of decay, is superior to chemical fertilizers containing equal amounts of potassium, phosphorus, and nitrogen compounds".

In 1918, Throckmorton and Call warned that although commercial fertilizers are more concentrated, they do not supply organic matter, which is "absolutely necessary to supply plant food, to preserve good tilth, and to retain water in the soil." Because commercial fertilizers do not supply organic matter, "they can't be expected to replace manure in soil improvement, but should be used, where they can be used profitably, in addition to barnyard manure and other forms of organic matter". This caution continued for more than thirty years while Kansas farmers were advised to use legumes and manures as nitrogen sources and pay careful attention to nutrient cycling on their farms. In 1956, the Kansas State Agricultural College publication, "Legumes vs. Commercial Fertilizer" documented a major change on Kansas farms. The authors reported on a study of the economics of Kansas cropping systems. One objective was to understand why farmers were planting fewer legume acres than were recommended. "Data indicate considerable advantage to certain legume rotations...Even though this is more profitable, farmers probably grow fewer legumes because they need income quickly".

Once a farmer began using commercial fertilizers, the soil would be slow to return to a more natural cycle of fertility that did not include commercial fertilization. In 1918, farmers were already asking whether commercial fertilizers "impoverish the soil." There were reports that farmers who stopped using commercial fertilizers experienced reduced crop yields. Throckmorton and Call explained that these fertilizers "cannot in themselves be expected to maintain the fertility of the soil. They should, therefore, be used only when a good rotation of crops is practiced, and when organic matter is supplied systematically".

Crop Rotations

In the 1910 bulletin, "Fertilizers and Their Use," the authors speculate that there is something other than the chemical analysis of the soil, which affects plant growth. They expected that rotations of crops may have an impact on "soil conditions". Without fully understanding the dynamics of rotations, farmers and researchers already considered them an essential part of good soil management.

By 1935, researchers were refining their view of the use of rotations. "Rotation of crops should not be loosely recommended without stating specifically what the rotation should be, or having in mind the wide differences existing between possible rotations". Citing studies of soil fertility under various cropping patterns and soil treatments over a period of twenty years, Throckmorton and Duly emphasize that not all rotations can be used interchangeably.

Some rotations are less effective than others at maintaining soil fertility as well as at providing economic returns. In some instances, continuous cropping of hay or small grains for a few years was better for the soil or the pocketbook than rotations, which included row crops. Soil quality concerns aside, the prices of individual crops and their cost of production weigh heavily in determining the most profitable rotations. As prices fluctuate, the rotations providing the greatest economic return also vary. During the latter half of the 1930's, an analysis of these continuing soil fertility studies in Manhattan looked at nitrogen and organic carbon levels. One conclusion was that within the cropping patterns studied, "the larger the percentage of the crop cycle occupied by biennial or perennial legumes or sod crops, the higher will be this level (of nitrogen and carbon)." The use of manure, fertilizers, and lime, if they stimulated crop growth, particularly of a legume, would maintain higher levels of nitrogen and carbon.

Wrong Turns

Besides repeated references to the importance of organic matter in retaining soil moisture, researchers have addressed other means of moisture conservation. A very early bulletin from 1899 examined the possibility that fertilizers themselves might slow water evaporation from the soil. Various treatments were tried on outdoor plots and on small pots within the laboratory over a number of years. In every case, there was no difference in treatment.

Perhaps the most intriguing study was sponsored by the E.I. DuPont de Nomours Powder Company from 1911 to 1913. The dynamite industry had been heavily promoting the use of their product for improvement of all types of soils. The study included plots on heavy clay soils at the Fort Hays and Manhattan experiment stations as well as on numerous farms across the state. On a field eighty rods long and eight rods wide, thirty-inch holes were dug fifteen feet apart in rows sixteen feet apart. One half stick of dynamite was placed in each hole and exploded.

Data showed no significant differences in crop yields, soil moisture, nitrate levels, or bacterial activity on the dynamited soil. Unfortunately, the physical characteristics of the soil were considerably diminished. The explosion forced soil at the centre of the charge into the surrounding pore spaces producing a cavity surrounded by a hard, compact mass. These "jugs" would fill with water during rainstorms and hold it until evaporated.

The cost was prohibitive with the dynamite expense alone at $12.20/acre. Labour was an additional $5.00/acre. The researchers concluded, "In no instance was there improvement sufficient to pay the expense of dynamiting".

Erosion Control

Erosion was recognized as one of the factors resulting in soil depletion. Throckmorton and Call felt erosion could be prevented by deep plowing, adding organic matter and by "working the ground at right angles to the slope of the land". Although plowing would later be seen as a culprit in erosion, it may have offered an advantage over shallow disking by initially creating a rough surface more resistant to wind. The merits of organic matter in stabilizing the soil were universally touted throughout early publications. However, in 1957, researchers in western Kansas reported findings indicating that increases in the soil's organic carbon content did not necessarily mean less susceptibility to wind erosion. Heavy soils could be more vulnerable with the addition of well-decomposed organic matter. Undecomposed crop residues or other organic matter were beneficial in slowing the effects of the wind.

Farmers could create another tool to control wind erosion by alternating strips of crops. Researchers in the 50's set out to determine the ideal width of these strips for maximum protection. They considered the quantity of crop residue and soil roughness that would be produced during years of low rainfall, high wind, and low crop yields. Full wind protection during such years required strips so narrow that they would be impractical to farm. When combined with other erosion control methods such as high residue management, the field strips could be increased in size to an acceptable level and still provide a high degree of protection. A 1962 bulletin on farming systems in western Kansas notes that contour farming resulted in increased yields for a wheat, wheat, sorghum, and barley rotation at Fort Hays. These researchers also experimented with dikes around very level fields to capture moisture and found some yield advantage.

Western Kansas Cropping Systems

Kansas State research bulletins and publications documenting the work in western Kansas over the first half of the 1900's reflect a very different type of cropping system from that used in the eastern parts of the state. Dryland farming in areas of low rainfall and high winds presents special challenges for soil health. This flat, arid region in the western half of the state supports a fragile wealth that requires careful consideration to maintain soil resources.

Soil studies in western Kansas date back to the very earliest years of the twentieth century. A continuous study of organic carbon and nitrogen in soils at Ft. Hays Experiment Station lasted for more than thirty years with follow-up research continuing for at least another fifteen years. The western Kansas studies covered a number of topics including the use of fallow periods, changes in organic carbon and nitrogen, tillage methods, and crop rotations. In nearly every instance, the principle concerns were the conservation of soil moisture and fertility.

Fallow

Fallow is the "practice of keeping land free of all vegetation throughout one season for the purpose of storing moisture for a crop the following year. Where rainfall is too low to support yearly crop production, fallow can be an important part of the cropping system.

Using fallow to store soil moisture results in production stability by decreasing the number of crop failures. Adherence to the fallowing pattern in high rainfall years is important since a portion of that moisture is stored for following dry years.

In 1962, Ft. Hays researchers reported that milo yields after fallow were twice the yields of milo crops following milo. Although this was the greatest percentage increase for any crop studied, all crops saw increases (1962, Investigations of Cropping Systems, Tillage Methods, and Cultural Practices for Dryland Farming). A 1941 publication states that corn, oats, and barley grown on fallowed ground show a marked increase in quality even if the yield from one year does not equal two crops. Fallow in a rotation with forage crops "allows farms without pasture to reintegrate livestock and supplement the carrying capacity of those farms with pasture".

The successful use of fallow is related to soil type with its greatest value seen on heavier soils that have an increased moisture storage capacity. Light soils without a heavy subsoil, shallow soils, and hilly topography are not likely to provide an economic advantage with fallowing since moisture cannot be held.

Soil management techniques for fallow rotations must always be directed toward capturing moisture, preventing evaporative losses, and timely destruction of weeds. The average rainfall in western Kansas may be adequate to produce a crop however the moisture losses from evaporation, runoff, and weed pressure rob a large portion of the water before it can be utilized. Throckmorton and Myers hypothesize that the farmer has little control over evaporative losses which can be 60-75% of total precipitation. "Fallow is extravagant in so far as storage of total precipitation is concerned, but it is essential as a means of stabilizing production through having sufficient moisture in the soil at seeding time to justify the seeding of a crop".

Tillage is a key to the successful management of runoff and weed growth. "A good summer fallow is one in which the soil is free of all growing plants throughout the fallow period and has a rough open surface which will permit a ready and rapid penetration of moisture." If possible the stubble of the preceding crop should be left standing during the winter and spring to capture snow and prevent wind erosion. Thereafter, cultivation should be used when weeds are still small and the soil will form clods to create a rough surface. To further protect the soil from wind erosion, fallow strips can be alternated with crop strips following field contours. The fallow that is poorly managed is doubly exposed to the wind. Proper management is an effective means of checking erosion.

In 1962, Luebs adds that shallow cultivation using a one-way disk plow or a subsurface tillage tool is just as effective as a plow or a lister for destroying weeds. This type of tillage leaves plant residues on the surface of the soil to slow the wind action.

Organic Matter Content of Western Kansas Soils

General soil management techniques would indicate that increasing the organic matter content of the soil would be one tool to build its water holding capacity. Long term studies of organic carbon and nitrogen levels in soils at Colby, Hays, and Garden City provide an interesting look at organic carbon levels.

All cropping systems examined in the studies were resulting in decreases in both soil carbon and nitrogen levels. The cropping systems were generally depending upon native fertility of the soil or some applications of manure for crop production. Continuous small grain production or small grains in rotation with a fallow period showed the least destruction of soil organic matter levels. The practice of fallowing in itself decreased organic matter levels since nothing was allowed to grow on the soil but when fallow was used in a rotation with small grain crops, the losses were decreased.

The use of green manure crops to build organic matter might be considered in a higher rainfall area. However, as early as 1918, researchers warned western Kansas farmers that green manure crops would use too much moisture prior to the main crop. They recommended finding some other source of organic matter.

Sampling for the long-term studies began in 1916 and in each successive report, researchers found that applications of animal manure and/or straw slowed the loss of organic carbon and nitrogen or provided a small increase. The relationship with regard to crop yields was less positive.

In 1943, researchers reported that manure applications increased yields only under "certain conditions" and added that "perhaps (manure's) advantage will become more evident in the future. In 1957 manure and straw applications showed no benefit to yield data. Citing the maintenance of nitrogen and organic matter content, researchers concluded that "these results indicate, even at the present time, that all manure should be conserved and applied to the land.

In 1962, researchers stated the manure was of "negligible" value for wheat and sorghum in a fallow-wheat-sorghum rotation. In 1965, after more than forty years of soil studies at the Ft. Hays Experiment Field, the researchers acknowledge that green manures and animal manure applications lower nitrogen and organic carbon losses. But they maintained that their use to maintain soil productivity "in this area is not practical" citing the need for 25-30 tons of manure per acre every three years to maintain nitrogen and carbon levels.

The data regarding crop yields and carbon levels in these studies did not vary significantly over the course of forty years. However the conclusions regarding "practicality" are considerably different. It may be that as farming practices changed and manure was less accessible on the average farm, its use was, indeed, less practical. As livestock concentration has increased in some parts of western Kansas at the end of the twentieth century, we are once again revisiting the practicality of manure applied to cropland. Although the economics may be driven by a need to dispose of excess manure, wise use indicates the "practicality" of using those manures to build fertility and organic matter levels.

Fertility of Western Kansas Soils

"From a practical standpoint, the question of the use of nitrogen fertilizers in western Kansas presents a number of important problems." Researchers in 1943 could see that frequent crop failures and low yields were due to limiting factors other than fertility - principally moisture. They also felt that the use of summer fallow "will reduce the need (for fertilizers) due to the accumulation of nitrate nitrogen". For years they examined nitrogen changes in the soils and considered the best course for maintaining fertility at a level that permitted crop production with an economic return.

The first sixteen years of the long term soil studies at Hays, Colby, and Garden City showed nitrogen losses from 1916-1938 that were nearly equal to the nitrogen removed by the crops. The losses immediately following sod breaking were greater than the losses later in the study. For a time, there was hope that this trend indicated a possible equilibrium for nitrogen levels.

Nitrogen lost through crop removal was being somewhat offset by the deposition of nitrogen in rain and snow during the fallow period. However, researchers estimated this amount to be only three to eight pounds per acre per year with an average deposition of 3.44 pounds per acre each year. They also speculated that there might be some fixation of nitrogen by free-living bacteria (Azotobacter and Clostridium) although they had not been able to verify this was true in any field experiments. Crop yields in these early studies fluctuated so much depending on rainfall patterns, no trend in decreasing yields due to fertility could be tracked. It is most likely that farmers were still mining the native fertility of the soil but at a much slower rate than the farmers in the eastern half of the state due to the difference in rainfall. Considering the fragility of western Kansas cropping systems, farmers in that area need to be able to adapt their cropping patterns to fit the current conditions. "The successful dryland farmer in this area must, as far as possible, be flexible in choosing cropping sequences, tillage methods, and cultural practices". He or she must consider the weather, soil conditions, preceding crops, residue, weed populations, soil moisture, tilth, and the removal of nutrients.

Looking to the Future of Kansas Soils

Although healthy soil is the basis for the rich diversity of agriculture in Kansas, a host of barriers, both economic and social, prevent us from protecting and building soil quality. Short term leases of rented farm ground discourage the use of practices that only see an economic return after multiple years. Inexpensive commercial fertilizers have precluded the need to monitor nutrient cycling on the farm. New interest in protecting water quality, conserving water use, managing excess livestock wastes, and developing farming systems that reduce the use of commercial fertilizers and pesticides may refocus attention on some of the same questions that were addressed during the first half of the twentieth century. Farmers may gain understanding from the early research of Kansas State University and then begin to ask the questions that will lead us into the twenty-first century with renewed interest in our soil.

Role of Bacteria in Coffee Plantation Ecology

To date we have written a dozen or more articles on the pro active role of microorganisms in shaping the destiny of the Coffee Mountain. However, off late we have received hundreds of e. mail requests from all over the globe, asking us to elaborate on the type of microorganisms carrying out these functions. Armed With a back ground of Microbiology and Horticulture, we have made an honest attempt to simplify the role of BACTERIA in coffee plantation ecology. All living organisms are classified as either PROKARYOTIC (PRIMITIVE) or EUKARYOTIC (Higher forms of life) based on their cellular structure. Bacteria and blue green algae are grouped under prokaryotes and all other organisms are eukaryotes. Bacteria are unicellular microscopic or single celled organisms widely distributed in nature. The greatest benefit in studying bacteria is that it throws light on the evolution of simple cellular systems and the higher forms of life.

Microorganisms are categorized into six distinct groups.

- Bacteria
- Fungi
- Actinomycetes
- Algae
- Protozoa
- Viruses.

The primary purpose of writing this article is to help coffee farmers worldwide in understanding the basic premise on which BACTERIA operate and the ways and means of carefully exploiting their potential to maintain a perfect ecological balance within the coffee farm.

The coffee habitat provides a fertile ground generating a host of both macro and micro organisms. The bacteria are the most dominant group of microorganisms in coffee soils. The microscopic analysis of coffee soils shows that there is plenty of room at the bottom for the proliferation of different types of microorganisms because of the rich humus and organic matter content. Among the different groups, bacteria are capable of harvesting atmospheric nitrogen, solubilisation of rock phosphate and in the transformations of various substrates resulting in periodic supply of available nutrients for plant growth and development. However, coffee farmers need to understand that the types of bacteria and their numbers are governed by the soil type and cultivation practices like addition of chemical manures, pesticides, poisons, type of tillage etc. In turn the activity of bacteria is influenced by the availability of nutrients, both in organic and inorganic forms. Their numbers are very high in coffee soils with large number of trees compared to open meadows. This is due to the shading nature as well as greater root density and the abundant availability of soil organic matter. Due to the high organic matter content of coffee soils, bacteria decompose the organic matter and in the process acquire energy.

Bacterial cells are so very small that when you think of these fastidious and ubiquitous microbes, you need to think small. They are measured in microns and the equivalent of one micron is: one by thousandth of a millimetre. Hence one needs a fairly high powered microscope to observe these minute wonders. However, they have a remarkable advantage because they have their strength in numbers. The population of bacterial cells in soils is always great. Due to their rapid growth and short generation time they can quickly act on various organic materials. In harsh environments lacking oxygen the bacteria alone are responsible for almost all the biological and chemical changes. Because of their very small size bacteria have a very high ratio of surface area to volume. Also, since bacteria are single celled microorganisms they absorb their nutrients through their cell membrane, there by exhibiting very high metabolic rates.

The earliest inhabitants of Planet Earth have undoubtedly been the microorganisms. From primitive prokaryotic unicellular microorganisms, evolved the higher forms of life. Microorganisms have thus been the earliest participants in shaping various life processes. Microorganisms have been largely responsible in changing the primordial atmosphere resulting in the formation of gaseous oxygen needed for plant growth. Fifteen million years of evolution has shaped the coffee forest and today their future is in our hands. Fundamentally, it has been an evolution of skills.

The evolutionary ladder points out to the pivotal role played by microorganisms in adapting to harsh environmental conditions, formation of tripartite bonds, break down of complex polysaccharides into simpler molecules and in the process providing the energy needs of the biotic community. It is in this context that this article throws light on the role of microorganisms, starting with BACTERIA in transforming the coffee landscape into an evergreen rich forest. Among the different microorganisms, Bacteria are known to play a vital role in the distribution and supply of energy needs of the entire coffee mountain. Decomposition of almost all insoluble salts is mediated by one or the other group of bacterial communities.

Distribution and Functions of Bacteria

Bacteria are widely distributed along the length and breadth of the coffee mountain. In short they are found almost every where. Bacteria are single celled organisms and in spite of their simplicity are highly efficient. Their numbers decline with depth of soil.

A majority of the coffee farmers are unaware that the great majority of bacteria are beneficial and absolutely necessary to convert farm wastes, organic debris and other by products into energy rich compounds needed for plant growth and development. Plants and animals depend on the fertility status of the soil and this in turn is dependent on the activity of soil microorganisms. Plants cannot directly utilize organic compounds such as fatty acids, lipids, carbohydrates and proteins. Microorganisms are a vital link in the mineralization of organic constituents and provide nutrients in the available form for plant growth and development. Bacterial cells can withstand long periods of drought due to the protective cover around the cell wall known as CAPSULE. The capsule is a slimy or a gelatinous material and encloses either one cell or a group of cells. At times the bacteria make use of the polysaccharides present in the capsule as a source of reserve food material. The capsule enables the bacteria to avoid predation by larger soil microbes and infection from viral strains. In addition to protection, capsules also play an important role in the attachment of bacterial cells to plant or rock surfaces and in the formation of biofilms.

Bacteria are morphologically grouped into three types. Cell structure is a key element in the characterization of bacteria.

1. Cylindrical or Rod shaped commonly referred to as Bacilli. They are the most numerous. Bacillus species are known to overcome extreme weather conditions by the formation of endospores that function as part of the normal life cycle of the bacterium. These endospores are resistant to long periods of drought and desiccation. With the on set of favorable conditions the spore germinates and a new bacterial cell grows.

2. Communication between all life is essential in unfolding the various patterns of life. Without the ability to communicate, life, including the simplest single celled organism, could not exist. Bacteria have the capacity to analyze vibrations in the surroundings and accordingly react. Besides shape and size certain rod shaped bacteria have thin hair like appendages on the outer cell wall known as flagella, which can sense the external environment and constantly send out chemical signals to reach out to other communities. Flagella are believed to be organs of locomotion.

3. Spherical or ellipsoidal bacteria are called cocci. Ellipsoidal bacteria occur in pairs and are referred to as STREPTOCOCCI,

when in four cells, arranged in a square they are known as TETRADS; when in irregular clusters like a bunch of grapes they are called STAPHYLOCOCCI and when arranged in a cubical form known as SARCINAE.

4. *Spiral or Helicoidal*: Winogradsky a leading soil microbiologist placed Soil bacteria into two broad divisions.

Autochthonous Species

These refer to the indigenous or native species. The population of these bacteria is always uniform and constant in coffee soils because their nutrition is dependent on the native soil organic matter.

They multiply rapidly in the presence of large quantities of biomass, organic matter, humus, and other soil amendments having a low C:N ratio. They are pretty tough and resistant to varied agro climatic conditions. They participate in all biochemical functions of the community. The presence of these bacteria is fairly high and their numbers are constant. The presence or absence of specific nutrients does not change their numbers significantly.

Allochthonous Species or Zymogenous Bacteria or Fermentative: Commonly referred to as the invaders. Their participation in biochemical functions is insignificant.

These bacteria are active fomenters and need nutrients which are quickly exhausted. They are involved in a process in which organic matter is rapidly attacked in successive stages and made available to the plants.

At each stage of decomposition a specific group of organism is involved. The bacterial numbers increase rapidly whenever furnished with the special nutrients (leaf litter, biomass, compost) to which they are adapted. On exhaustion of these nutrients their numbers decrease and return with the addition of nutrients. Hence, this group of bacteria requires an external source of energy for their multiplication and growth.

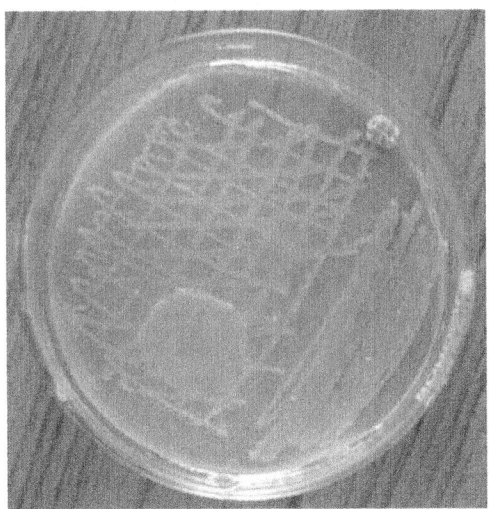

Bacteria in this group include the nitrogen fixers, phosphorus solubilisers, nitrifiers, cellulose hydrolyzing bacteria, sulphur oxidizers, spore forming bacillus and non spore forming pseudomonas.

Environmental Factors

Bacterial numbers, their density, type and composition is governed by the environmental Stimulus.

Aeration

Bacteria are further divided as

Aerobes: Require the presence of oxygen for growth and metabolic activity.

Anaerobes: Bacteria which grow in the absence of oxygen.

Facultative Anaerobes: Develop either in the presence or absence of oxygen.

Aerotolerant Anaerobes: These bacteria grow under both aerobic and anaerobic conditions.

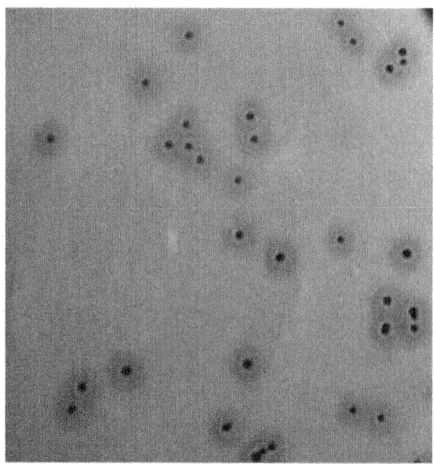

Moisture

Aerobic bacteria are the main stay in coffee soils and the optimum level of moisture content for their activities is in the range of 50 to 75% of the soil's moisture holding capacity. Coffee soils are inherently shaded by tree canopies as well as by the coffee bush. Hence they remain shaded most of the time. Also, a host of factors result in the availability of moisture throughout the year. For e.g. The south west and the north east monsoon together keep the soil moist for eight months of the year and the remainder months, due to soil conservation practices adopted by the Indian coffee farmer , the moisture is always available for bacterial growth and development.

Water makes up a major component of the microbial cell. Hence it is a key component for the functioning of the cell. The most common problem encountered in coffee soils is not the lack of moisture but the availability of excess moisture which is detrimental for the growth and multiplication of bacteria. Excess moisture limits the supply of gaseous oxygen resulting in an anaerobic environment. Water logging brings about a decrease in the abundance of bacteria.

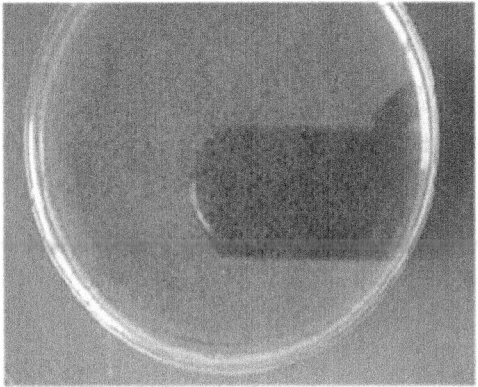

Temperature

Bacteria are highly sensitive to temperature fluctuations. Apart from growth and development, temperature plays a vital role in the biochemical processes carried out by the bacterial cell.

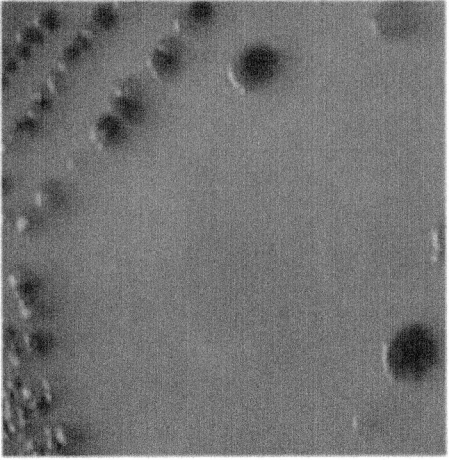

Mesophiles are the ones which grow well in the temperature range of 25 to 35 degree centigrade. These constitute the bulk of the coffee soil bacteria. For most part of the year the temperature profile in the coffee mountain falls in this range. Some scientists have further divided mesophiles as :

Oikophilic; Organisms whose optimum temperature is around 20 degree centigrade. Somatophilic; Organisms whose optimum temperature is about 37 degree centigrade.

Psychrophyles are bacteria that love cold and grow at temperatures below 20 degree centigrade. Thermophiles are temperature loving bacteria and grow best in the temperature range of 45 to 65 degree centigrade. These bacteria are active in compost pits.

Organic Matter

The population of bacteria is directly related to the organic matter content of the soil. Due to periodic leaf shedding and availability of huge quantities of carbonaceous materials on the floor of the coffee forest, the bacterial numbers is the largest. Also the coffee farmers incorporate green manures, compost and biomass from time to time which act as stimulants for the growth and proliferation of bacteria.

Acidity

The optimum pH for the growth of bacteria is NEUTRAL pH. Coffee farmers need to keep the hydrogen ion concentration of their soils close to neutral because in highly alkaline or highly acidic conditions the growth and multiplication of bacteria is inhibited. In general in heavy rainfall areas receiving 100 inches and more it is advisable to apply lime or dolomite once every two years and in moderate rainfall regions, once every four years. This practice will not only increase the bacterial numbers but will also enable the coffee bush to take up inorganic nutrients in a more efficient way.

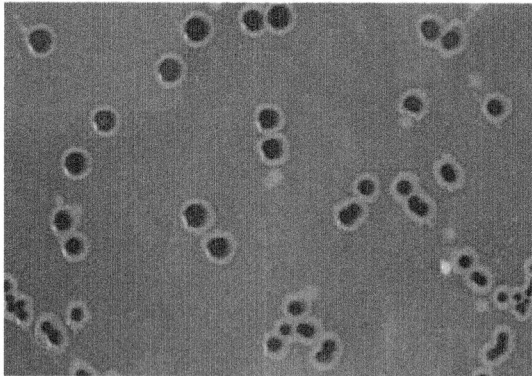

Figure : Acidophilic Bacteria: Bacteria capable of growth in extremely low Ph.

Alkalophilic Bacteria: Bacteria capable of growth in extremely alkaline soils (p H 10.5)

Halophilic Bacteria: Bacteria capable of tolerating high salt concentrations.

Xerophilic Bacteria: Bacteria capable of growth in dry habitats.

Inorganic Nutrients

Application of fertilizers and chemicals greatly affects the bacterial population. Coffee farmer's world wide use ammonium fertilizers as the bulk of fertilizer application. Coffee farmers do not realize that ammonium fertilizers tend to lower the soil pH resulting in acidity due to the microbial oxidation of ammonium to nitric acid. More than the effect of fertilizer, it is the acidity which suppresses the bacterial population. This problem can be easily overcome by split applications spread out over a two week period. More importantly, the application of fertilizer should be carried out when the soil moisture is optimum. It is a proven fact that small amounts of inorganic fertilizers supply the needs of the bacterial community in the form of inorganic nutrients.

Farm Practices

Farm practices also exert direct and indirect biological effects on the coffee farm. Periodic soil disturbance will affect the bacterial population. Addition of organic manures from time to time and incorporating legumes into the soil with proper carbon nitrogen ratio accelerates the build up of beneficial micro flora. However, if soil hardens up over a period of time, then it will have an adverse effect on the bacterial numbers.

Nutritional Requirements of Bacteria: Macronutrients

Carbohydrarates, Proteins, Lipids, Nucleic Acids. Carbon requirement is the greatest, followed by nitrogen, phosphorus and sulphur. Potassium, sodium, calcium and magnesium are also required in substantial quantities.

Micronutrients

Cobalt, iron, zinc, copper, molybdenum, manganese.

Certain bacteria require specific organic compounds that they are unable to synthesize from simple compounds. Hence they require GROWTH FACTORS classified into one of the following groups; Amino Acids; Purines & Pyrimidines; Vitamins.

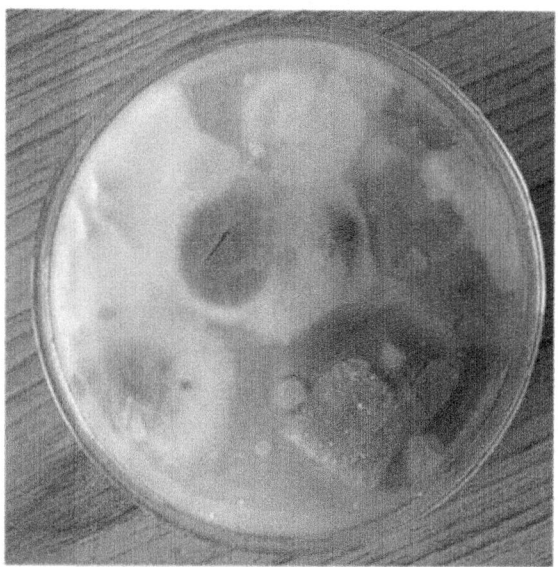

The dominant groups of bacteria are the heterotrophs but a few genera have photo chromatic pigments enabling them to have photoautotrophic nutrition. Two main classes of bacteria are the Heterotrophs or Chemoorganotrophic: Bacteria which require preformed an organic nutrient which serves as a source of energy and carbon. Autotrophic or Lithotrophic: Bacteria which obtain their energy from sunlight or by the oxidation of inorganic compounds and their carbon by the assimilation of carbon dioxide. Autotrophs are further classified as Photoautotrophs or Photolithotrophs; where energy is derived from sunlight and Chemoautotrophs or Chemolithotrophs which obtain their energy from the oxidation of inorganic materials.

Specialised Bacterial Cells

Endospores: Bacterial communities have developed their own specialized skills to survive the hardships of nature. At times it involves a constant battle where it is not only the survival of the fittest but survival by way of forming alliances with other biotic communities. It involves constant signal exchange between predators and prey, struggles for dominance, defence of territories and many ways to simply survive by the production of spores and endospores which are tolerant to adverse weather conditions.

These structures are resistant to heat, desiccation, high salt concentrations, cold, osmosis and chemicals, compared to the vegetative cells producing them. Endospores are bodies produced within the cells of a considerable number of bacterial species. Sporulation confers protection to the cell whenever the occasion arises. Because of their low rate of metabolism , endospores can survive for a number of years without a source of nutrients.

However, when favorable conditions appear, endospores begin to germinate within a few minutes to form a new vegetative cell.

Most Commonly Encountered Soil Bacteria

Belong to the Following Genera

Aerobacter, Myxobacteria, Bacillus, Pseudomonas, Flavobacterium, Arthrobacter, Achromobacter, Clostridium, Corynebacterium, Mycobacterium, Sarcina, Myxococcus, Archangium, Chondrococcus, Cytophaga, Sporocytophaga, Polyangium.

Figure : Chemical Composition of the Bacterial Cell on Dry Weight Basis.

The major constituent is water to the extent of 90 %. The ash content varies. Carbon 45-55% Nitrogen 8-15%. The ash from bacteria may contain Phosphorus 10-50%, Potassium 4-25%, Sodium 10- 35%, Magnesium 0.1-10%, Silica 0.5-7.7%, Calcium 0.3-14%, Chlorine 1-44%, and trace amounts of iron. These variations are due to the different bacterial species on the floor of the coffee mountain.

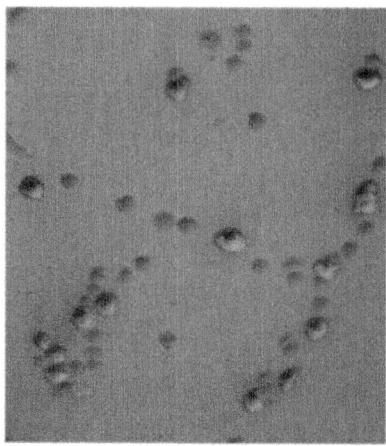

Isolation

We have had the rare opportunity of isolating thousands of species of soil bacteria during our post graduate studies. Our study included soils from various agro climatic regions of Karnataka. Our results had far reaching consequences. We were able to prove that soils with neutral pH harboured the maximum load of nitrogen fixers and phosphate solubilisers. Different soil types had different species of bacteria but most of them were beneficial and acted as important links in various soil transformations.

Conclusion

Shade grown Indian coffee plantations have sustained many generations of farmers because of their resilience in overcoming all odds. The secret behind this success is attributed to the microbial inhabitants. Microorganisms cannot be seen by the naked eye, yet they constitute about one quarter of the biomass-the total weight of living organisms in the world. Animals and plants account for the remainder. In the strict sense, more than 98 % of what we describe as waste inside the farm is valuable food for one or the other group of microorganisms. Bacteria recycle these wastes into power packed energy rich nutrients required for the survival of the coffee bush and its partners. Coffee farmers have an erroneous concept of the role of bacteria in nature. They strongly feel that the majority of bacteria are disease producing. We would like to set the record straight and state that the vast majority of bacteria are not only beneficial but are absolutely essential in building up a healthy coffee farm. Yes, there are a very few bacterial species that are harmful but in a healthy ecosystem they rarely express themselves.

It is very important that the coffee farmer understand the subtle role played by bacteria in the transformation of major elements like nitrogen, sulfur and phosphorus, biodegradation, neutralizing toxic wastes, bio-control agents and a host of other activities. Certain bacteria belonging to the families Thiorhodaceae , Chlorobacteriaceae and Athiorhodaceae contain bacteriochlorophyll and various carotenoids and are also capable of photosynthesis. Bacterial interactions with the coffee bush and the surrounding flora are known to improve plant growth and productivity. The fertility of the soil is directly dependent on the activity of soil microorganisms. The soil microorganisms mineralize insoluble and indiffusable organic constituents and make them available to plants. The common denominator to assess a healthy soil is the viable number of soil microorganisms. The shade grown Indian coffee ecosystem is unique in the true sense that it simultaneously achieves the goals of agricultural production in terms of coffee, pepper, citrus, vanilla, areca nut and cardamom production on one hand and the conservation of biodiversity on the other hand. Nothing would be wiser for the world's coffee producing Nation's to follow in India's footsteps and grow coffee under the canopy of trees. There in lies the path to sustainability. In our humble opinion we strongly feel that only shared prosperity can make the future of this planet secure. At JOE'S Sustainable farm we work with ideas that might work or might not work. But in the end analysis, these ideas are the core to the survival of Planet Earth.

Bibliography

Adler-Nissen, J. : *Enzymic Hydrolysis of Food Proteins,* London: Elsevier Applied Science, 1985.

Alfred Byrd Graf: *Advances in Plant Physiology*, Rajat Pub, Delhi, 2008.

Allchin, F.R.: *The Agriculture of Early Historic South Asia*, Cambridge University Press, U. K., 1995.

Bagdi, Madan Lal : *Physiology, Biochemistry and Biotechnology,* Manglam Pub, Delhi, 2007.

Bora, K.K. : *Agros Dictionary of Plant Physiology and Biochemistry,* Agrobios, Delhi, 2001.

Bose, Bandana and A. Hemantaranjan : *Developments in Physiology, Biochemistry and Molecular Biology of Plants,* New India Pub, Delhi, 2005.

Boyer , J.S.: *Measuring the Water Status of Plants and Soils,* Academic Press, N.Y., 1995.

Chadha K. L. and Pareek O. P.: *Advances in Horticulture: Fruit Crops,* New Delhi, Malhotra Publishing House, 1993.

Chakraborty, Chiranjib : *Advances in Biochemistry and Biotechnology,* Daya, Delhi, 2005.

Chaudhary, Vikas : *Entomology and Pest Management,* Navyug, Delhi, 2008.

Chopra, P.N.: *Agriculture of Ladakh: Our Cultural Fabric Series.* New Delhi: Ministry of Education and Social Welfare, 1978.

Clark, Vijay Paul: *Physiology of Crop Production,* International Book, Delhi, 2007.

Collymore L.: *Fruit Production in Barbados,* Port of Spain, Trinidad and Tobago, 1996.

Coste R.: *Coffee: the Plant and the Product,* London, MacMillan, 1992.

Coultate, T. P. : *Food: The Chemistry of its Components,* London: The Royal Society of Chemistry, 1988.

Dabholkar, A.R. : *General Plant Breeding,* Concept, Delhi, 2006.

Dar, Ghulam Hassan : *Soil Microbiology and Biochemistry,* New India Publishing Agency, Delhi, 2010.

Devi, C.R. Sudharmai : *Analytical Procedures in Soil Science and Agricultural Chemistry*, Agrotech, Delhi, 2004.

Dharamvir Hota: *Modern Biotechnology in Plant Breeding*, Gene Tech Books, Delhi, 2007.

Doijode S. D.: *Seed Germination in Fruits*, New Delhi, Malhotra Publishers, 1993.

Dutta, Naren Kumar : *Fundamentals of Biochemistry : A Practical Approach*, Kanishka, Delhi, 2005.

Featherly H. I.: *Taxonomic Terminology of the Higher Plants*, USA, Iowa State College Press, 1954.

Ghosh, A.R. : *Biopesticide and Integrated Pest Management*, APH, Delhi, 2009.

Goel, A K : *Basic Concept of Organic Chemistry*, Pearl Books, Delhi, 2008.

Hardy B.: *Biology and Agronomy of Forage Arachis*, Cali, International Centre for Tropical Agriculture, 1994.

Herminie Broedel Kitchen: *Soils and Crops : Diagnostic Techniques*, Satish Serial Publishing, Allahabad, 2004.

Jasra, O.P. : *Environmental Biochemistry*, Sarup & Sons, Delhi, 2002.

John C. Benghin: *Global Agriculture Trade and Developing Countries*, Manas, Delhi, 2005.

Jones, R. M.: *Plant Resources of South-East Asia,* Wageningen, Pudoc Scientific Publishers, 1992.

Kannaiyan, S. : *Rice Management Biotechnology*, Associated, Delhi, 1995.

Kapoor, R.L. and M.L. Saini: *Plant Breeding and Crop Improvement*, CBS, Delhi, 1997.

Kataria, T N : *Plant and Crop Physiology*, Pearl Books, Delhi, 2008.

Kaul, M.K. : *Medicinal Plants of Kashmir and Ladakh : Temperate and Cold Arid Himalaya*, Indus, Delhi, 1997.

Khan, Samiullah: *Plant Breeding Advances and in vitro Culture*, CBS, Delhi, 1997.

Khanna, V K : *Objective Genetics, Biotechnology, Biochemistry and Forestry*, I.K. International Publishing House, Delhi, 2008.

Kumar, Arvind : *Environmental Pollution and Agriculture*, APH, Delhi, 2002.

Kumar, Ravi Theodore: *Farm Diversification for Sustainable Agriculture*, International Book Distributor, Delhi, 2002.

Kumar, T. Vijaya : *Financing Agriculture by Primary Agricultural Co-operatives Societies in India*, Madhav Books, Delhi, 2010.

Madan Lal Bagdi: *Physiology, Biochemistry and Biotechnology*, Manglam Pub, Delhi, 2007.

Mal, Pratap : *Infrastructural Development for Agriculture and Rural Development*, Mohit, Delhi, 2001.

McCabe, W., Smith, J. and Harriott, P. : *Unit Operations of Chemical Engineering*, 7th Edition, McGraw Hill, 2004.

Mistry, B.D. : *A Handbook of Spectroscopic Data : Chemistry*, Oxford University Book Co, Delhi, 2009.

Mohanty, A.K. : *Entrepreneurship in Agriculture : Scopes and Opportunities*, Agrotech Pub, Delhi, 2011.

Murty, S. : *Faces and Phases of Agriculture and Industry in India*, RBSA Pub, Delhi, 2004.

Murugesan, R. : *Energy Use Efficiency in Dryland Agriculture*, Kalpaz, Delhi, 2010.

Narasaiah, M. Lakshmi : *Financing of Agriculture by Regional Rural Banks*, Sonali Pub, Delhi, 2008.

Narvekar, Raghunath   : *Molecular Biochemistry : Principles and Practices*, Adhyayan Pub, Delhi, 2008.

Nobel, P. S.: *Physicochemical and Environmental Plant Physiology*, Academic Press, San Diego, 1999.

Ojha, Jai Shankar   : *Aquaculture and Nutrition and Biochemistry*, Agrotech, Delhi, 2006.

Pandey, Ashok : *Biotechnology : Food Fermentation (Microbiology, Biochemistry and Technology)* , Educational Publishers, Delhi, 1999.

Pareek O. P.: *Advances in Horticulture: Fruit Crops*, New Delhi, Malhotra Publishing House, 1993.

Parry M.L.: *Climatic Change, Agriculture and Settlements*, Dawson Folkestone UK, 1978.

Patel, S V and B A Golakiya : *DNA : A Bridge Between Biochemistry and Biotechnology*, New India Publishing Agency, Delhi, 2008.

Politycka, B and C L Goswami: *Plant Physiology : Research Methods*, Scientific, Delhi, 2007.

Prasad, S.K. : *Biochemistry of Carbohydrates*, Discovery Pub, Delhi, 2010.

Punja, Zamir K. : *Fungal Disease Resistance in Plants : Biochemistry, Molecular Biology and Genetic Engineering*, International, Delhi, 2005.

Rahman, H. and A.K. Mohanty: *Entrepreneurship in Agriculture : Scopes and Opportunities*, Agrotech Pub, Delhi, 2011.

Rai, Lajpat: *Experimental Designing and Data Analysis in Agriculture and Biology*, Agrotech, Delhi, 2010.

Seema Srivastava: *Plant Physiology and Biochemistry*, Campus Books, Delhi, 2009.

Sharma, J.P. : *Entrepreneurship in Livestock and Agriculture*, CBS Pub, Delhi, 2010.

Sharma, R K and S P S Sangha : *Basic Techniques in Biochemistry and Molecular Biology*, I K International, Delhi, 2009.

Sharma, Ramniwas : *Growth and Development of Agriculture*, Biotech, Delhi, 2006.

Sharma, Vijay Paul: *Glimpses of Indian Agriculture : Macro and Micro Aspects*, Academic Foundation, Delhi, 2008.

Shivanand Tolanur: *Practical Soil Science and Agricultural Chemistry*, International Book Distributing, Delhi, 2004.

Shubhrata R. Mishra: *Morphology of Plants*, Discovery, Delhi, 2004.

Singh, S K : *Biotechnology, Plant Propagation and Plant Breeding*, Campus Books, Delhi, 2008.

Sneh Lata: *Taxonomy I : Systematics and Morphology*, Sonali Publication, Delhi, 2005.

Sudharmai Devi: *Analytical Procedures in Soil Science and Agricultural Chemistry*, Agrotech, Delhi, 2004.

Swaminathan, MS : *From Green to Evergreen Revolution : Indian Agriculture: Performance and Challenges*, Academic Foundation, Delhi, 2010.

Swarnim, K. : *A Textbook of Biochemistry and Microbiology,* Surendra Pub, Delhi, 2010.

Trivedi, P C : *Plant Physiology : Current Trends*, Pointer, Delhi, 2007.

Tyagi, I.D. : *Plant Breeding and Genetics at a Glance*, South Asian, Delhi, 2005.

Vanangamudi, K. : *Principles and Methods of Plant Breeding*, International Book, Delhi, 2005.

Verma, L.R. and R.C. Sharma: *Diseases of Horticultural Crops: Fruits*, Indus, Delhi, 1999.

Vijay Paul Sharma: *Glimpses of Indian Agriculture : Macro and Micro Aspects*, Academic Foundation, Delhi, 2008.

Yadav, M : *Nutritional Biochemistry and Metabolism,* Arise Pub, Delhi, 2008.

Yoshida T.: *Cultivation of Citrus Genetic Resources for Evaluation of Characteristics*, Tokyo, JICA, 1996.

Index

A

Agricultural Microbes, 1.
Agricultural Pests, 62.
Animal Biosecurity, 209.
Antibiotic Resistance, 23, 24, 34, 35, 40, 141, 184, 195, 196, 197, 198, 201, 203, 205.
Antisense RNA, 94, 96, 98, 99.
Applications, 22, 23, 62, 75, 77, 82, 95, 121, 164, 220, 239, 244, 252, 265.
Approaches, 19, 23, 127, 141, 142, 144, 147, 148.
Aquatic Environments, 21, 49, 55.
Asexual Reproduction, 33, 49, 50, 56.

B

Bacteria, 1, 2, 3, 4, 6, 8, 11, 21, 22, 23, 24, 25, 26, 27, 28, 29, 30, 31, 32, 33, 34, 35, 36, 37, 38, 39, 40, 41, 42, 43, 44, 46, 57, 60, 62, 76, 77, 82, 86, 88, 94, 95, 100, 101, 102, 103, 105, 106, 115, 116, 119, 123, 124, 139, 141, 145, 150, 152, 161, 173, 174, 179, 195, 196, 197, 198, 199, 201, 202, 203, 204, 205, 207, 223, 224, 225, 226, 227, 230, 231, 232, 233, 234, 235, 238, 240, 242, 243, 244, 252, 253, 254, 255, 256, 257, 258, 259, 260, 261, 262, 263, 264, 265.
Bacterial Cells, 22, 24, 26, 27, 30, 82, 88, 138, 203, 254, 255.
Bacteriology, 24, 25, 37, 211.
Biological Hazard, 206.
Biological Warfare, 23, 150, 152, 153, 156, 157, 158, 159, 162, 163, 164, 165, 166, 167, 168, 169, 170, 171, 172, 178, 180, 189, 190, 212, 217, 218, 219, 220.
Biological Weapons, 150, 151, 154, 155, 156, 159, 160, 161, 163, 164, 165, 166, 167, 168, 169, 170, 171, 172, 173, 175, 177, 178, 179, 180, 181, 182, 189, 190, 191, 210, 213.

C

Cellular Structure, 6, 253.
Coffee Plantation Ecology, 252.
Communication, 255.
Crop Plants, 62, 80.

D

Distribution, 31, 47, 50, 197, 254.
DNA, 1, 4, 5, 6, 8, 9, 10, 12, 14, 18, 20, 21, 22, 23, 30, 34, 35, 36, 38, 41, 42, 43, 45, 46, 48, 54, 65, 68, 76,

78, 79, 80, 83, 84, 85, 86, 87, 88, 89, 90, 91, 92, 93, 96, 122, 125, 126, 127, 128, 129, 130, 131, 132, 133, 134, 135, 136, 137, 138, 139, 140, 145, 147, 198, 199, 205, 224, 225, 226, 227, 228.

DNA Technology, 76, 78, 80.

DNA Vaccination, 125, 126, 128, 131, 133, 136, 138.

E

Entomological Warfare, 171, 172, 217, 218, 219, 220, 221.

F

Frost Control Biotechnology, 94.

Fungal Diseases, 44.

Fungus, 44, 45, 47, 49, 51, 56, 57, 58, 60, 63, 65, 103, 106, 118, 119, 121, 156, 159, 228.

G

Gene Cloning, 87.

Gene Exchange, 88.

Gene Therapy, 23, 126.

Genetic Diversity, 1, 48.

Genetic Engineering, 20, 59, 79, 184, 205.

Genetic Modification, 59.

Green Revolution, 78, 79, 80.

H

HIV Vaccine, 141, 142, 143, 145, 148.

Human Disease, 15, 40.

I

Immune Response, 2, 19, 96, 125, 126, 128, 132, 133, 134, 135, 136, 138, 146, 149.

Insect Pests, 67, 104, 105, 106, 107, 108, 109, 111, 113, 115, 116, 119.

Insect Resistance, 118.

K

Kansas Soils, 236, 237, 238, 243.

M

Management, 76, 104, 105, 109, 119, 209, 230, 231, 236, 237, 246, 248, 249, 250.

Mediated Resistance, 96, 97, 99, 198.

Metabolic Engineering, 59.

Metabolism, 6, 10, 28, 31, 32, 34, 36, 41, 68, 121, 149, 203, 263.

Microbial Control, 118.

Microbial Pathogens, 103.

Microorganisms, 2, 24, 25, 31, 33, 38, 39, 44, 60, 62, 66, 75, 76, 77, 79, 82, 86, 93, 95, 100, 101, 104, 108, 116, 118, 119, 123, 156, 170, 198, 204, 227, 252, 253, 254, 255, 264, 265.

Molecular Biology, 41, 65, 83, 87.

N

Natural Populations, 66, 67, 70, 73, 74.

O

Operations, 160, 163, 173, 178, 189.

Organization, 17, 28, 165, 171, 191, 197, 202, 208, 213.

P

Plant Breeding, 79, 80.

Plant Materials, 242.

Polymerase Chain Reaction, 38.

Production, 3, 14, 19, 24, 29, 32, 33, 41, 42, 44, 51, 59, 63,

65, 72, 73, 78, 79, 80, 81,
82, 85, 86, 87, 95, 96,
108, 112, 134, 135, 138,
139, 145, 154, 163, 171,
175, 178, 179, 183, 191,
193, 194, 195, 197, 203,
209, 218, 223, 227, 228,
236, 244, 247, 254, 255,
264, 265.
Projects, 170, 203.
Property, 87, 110, 112.
Protein Synthesis, 11, 86.

R

Recombinant DNA, 76, 78, 79, 80,
87.
Research, 17, 23, 25, 54, 56, 60,
64, 81, 82, 93, 108, 127,
132, 136, 142, 144, 147,
149, 151, 160, 161, 162,
163, 164, 165, 166, 167,
168, 169, 171, 175, 178,
179, 181, 184, 185, 191,
192, 194, 197, 198, 202,
203, 205, 206, 207, 208,
210, 217, 218, 220, 221,
222, 228, 236, 237, 248,
252.
Resistant Pathogens, 199, 203.
RNA Interference, 18, 21, 22, 35.
RNA Strategy, 94.

S

Sexual Reproduction, 49, 50, 51,
52.

Soil Bacteria, 102, 232, 235, 244,
256, 259, 264.

T

Technology, 40, 76, 78, 79, 80,
87, 93, 94, 109, 155, 164,
185, 210.
Transgenic Animals, 82.
Transgenic Plants, 81, 96, 97, 98,
99, 119.

V

Vertebrate Immunity, 119, 123.
Viral Capsid Protein, 96.
Viral Movement Protein, 98.
Virus, 1, 2, 3, 4, 5, 6, 7, 8, 9,
10, 11, 12, 13, 14, 15, 16,
17, 18, 19, 20, 21, 23, 35,
81, 96, 97, 98, 99, 107,
117, 119, 122, 123, 126,
127, 131, 134, 135, 136,
137, 138, 140, 141, 142,
143, 144, 146, 147, 150,
152, 169, 170, 181, 191,
192, 193, 194, 206, 207,
208, 210, 214, 215, 216,
221.
Virus Diseases, 97.
Virus Resistance, 96.

W

Withstands Radiations, 223.

❑❑❑